Climate Change - Recent Observations

Edited by Terence Epule Epule

Published in London, United Kingdom

Climate Change - Recent Observations
http://dx.doi.org/10.5772/intechopen.100703
Edited by Terence Epule Epule

Contributors
Abdul Rehman Aslam, Fatima Farooq, Yosune Miquelajauregui, Adela Madariaga-Fregoso, Pallab Kumar Sarma, Nikhilesh Baruah, Rupam Borah, Rupshree Borah, Arunjyoti Sonowal, Rekhashree Kalita, Prasanta Neog, Prabal Saikia, Nipen Gogoi, Muhammad Amjad Bashir, Qurat-Ul-Ain Raza, Abdur Rehim, Hafiz Muhammad Ali Raza, Saba Ali, Muhammad Ijaz, Faiz Ul Hassan, Muhammad Umair Sial, Yucong Geng, Joel Audefroy, Sanata Traoré, Fassé Samaké, Amadou Hamadoun Babana, Eric Williams Cornelius, Gloria Ladjeh Essilfie, Mavis Agyeiwa Aacheampong, Salimatou Samaké, Olayinka C. Oloke, Nelson A. Akindele, Laura Ann Donkers, Abraham Cardenas - Tristan, Andres Gerardo Castro - Ovalle, Oscar Reyes - Cardenas, María Guadalupe Galindo Mendoza, Mozhgan Shirinpour, Ehsan Atazadeh, Ahmad Bybordi, Ashkboos Amini, Hassan Monirifar, Xubing Zhang, Wenxi Li, Liang Zhang, Kai Wang, Ruizhi Li, Terence Epule Epule, Simon Lafontaine, Martin Jemo, Driss Dhiba, Ayoub Kechchour, Soumia Achli, Lahcen Ousayd, Wiam Salih, Perez Lionnel Kemeni Kambiet, Vincent Poirier

First published in London, United Kingdom, 2023 by IntechOpen
IntechOpen is the global imprint of INTECHOPEN LIMITED, registered in England and Wales, registration number: 11086078, 5 Princes Gate Court, London, SW7 2QJ, United Kingdom

British Library Cataloguing-in-Publication Data
A catalogue record for this book is available from the British Library

Additional hard and PDF copies can be obtained from orders@intechopen.com

Climate Change - Recent Observations
Edited by Terence Epule Epule
p. cm.
Print ISBN 978-1-80355-498-3
Online ISBN 978-1-80355-499-0
eBook (PDF) ISBN 978-1-80355-500-3

For EU product safety concerns:
IN TECH d.o.o., Prolaz Marije Krucifikse Kozulić 3, 51000 Rijeka, Croatia,
info@intechopen.com or visit our website at intechopen.com.

Meet the editor

Prof. Terence Epule Epule is Professor of Agroclimate and Adaptation Sciences at Université du Québec en Abitibi-Témiscamingue (UQAT). Prior to this, he was Assistant Professor of Climate Change Adaptation at the International Water Research Institute (IWRI), Mohammed VI Polytechnic University (UM6P). He was previously an Associate Professor of International Farming Systems at UniLaSalle, France, and a postdoctoral scholar at McGill University and University of Montreal, Canada. His research is at the interface of climate, agriculture, and adaptations. He has published more than sixty-one peer-reviewed journal articles, two books, and seven book chapters.

Contents

Preface

As we all know, the veracity of global climate change is unequivocal. The recent Intergovernmental Panel on Climate Change AR6 report affirms this assertion, observing that global temperatures will continue to rise while precipitation will decrease in some parts of the world and increase in others. In fact, the impacts of climate change on the environment and the economy will not be exclusively negative. While most developing world countries will continue to exhibit varied levels of vulnerability due to low adaptive capacities, in Canada, for example, climate change might extend the growing season for several crops, thus enhancing food production.

This book project begins with an understanding that the impacts of climate change across the world are and will continue to be varied. The global scope of the contributions is intended to elicit global readership and provide an opportunity to have experiences from across the world in a single book.

Consequently, this book contains thirteen peer-reviewed chapters on varied but interconnected themes that are within recent observations in climate change scholarship.

The chapters are organized into four sections. Section 1 introduces the concept of climate change, examining current global perspectives and trends. Section 2 includes four chapters focusing on climate risk, resilience, and vulnerability. The chapter topics include the nexus of climate change, spatial and temporal monitoring, impacts of climate change on radiation, and climate risk. They provide varied perspectives on climate risks, sensitivity, and exposure. Section 3 discusses monitoring and assessment. These chapters underscore various options used by researchers to monitor the impacts of climate change. The four chapters in this section address genetic interactions, microbial and chemical manipulation, trends in the use of fertilizers, and program evaluations. Finally, Section 4 focuses on adaptation. The four chapters in this section examine aspects of adaptation and mitigation of climate change, knowledge co-creation as an adaptation strategy, bio-stimulants and vulnerability to climate change, and the potential for domestic adaptation.

The scope of this book is to provide a niche not only for experts in climate change scholarship but also for other connected disciplines that are impacted by climate change ranging from activities that can be classified as primary, secondary, or tertiary. A broad readership including academics, students, policymakers, and various organizations and stakeholders looking for a place to obtain a broad perspective on climate change will find what they are looking for in this book.

This work has been made possible thanks to IntechOpen and the authors who provided their manuscripts. I would like to thank our team at Unité de recherche et développement en agroalimentaire (Unit of Research and Development in Agri-Food) (URDAAT), Université du Québec en Abitibi-Témiscamingue (UQAT) for encouragement and for providing the environment in which this work was carried out.

Terence Epule Epule
Professor of Agroclimate and Adaptaion,
Unit of Research and Development in Agri-Food (URDAAT),
University of Quebec in Abitibi-Témiscamingue (UQAT),
Notre Dame du Nord, Québec, Canada

Section 1

Introduction

Chapter 1

Introductory Chapter: Climate Change

Terence Epule Epule, Vincent Poirier and Wiam Salih

1. Introduction

Global evidence of climate change is unequivocal. This conclusion is based on recurrent observations of variations in numerous climatic variables. This phenomenon is further characterised by a paradox as both the direction of the changes in climate variables and effects often move towards opposite directions [1–3]. This is seen as while precipitation is increasing in some parts of the world, climate change is at the same time driving declining precipitation in other parts of the world. Across Africa, for example, the recent Intergovernmental Panel on Climate Change (IPCC) AR6 argues that for various global warming scenarios (1.5° C, 2° C and 4° C), northern and southern Africa will continue to witness declining precipitation and increasing temperature while west, middle and east Africa will witness increasing temperatures and increasing, sporadic but unreliable precipitation [4].

As can be seen in the scenarios above, various global warming scenarios are driving climate change. Even though the evidence of climate change is unequivocal, the scientific literature has been somehow divided on the causes of climate change [5]. In this case, three major schools of thought emerged. The first school of thought argues that global warming, for example, is driven by mainly climatic or biophysical drivers and while the second posits that non-climatic drivers are mainly responsible for global warming [6]. The third school of thought argues that both climatic and non-climatic drivers have a role to play in the occurrence of global warming and, consequently, climate change. Whichever direction the debate is going, both climatic and non-climatic forcing have a critical role to play, as discussed by [6, 7]. If the global rise in temperature is at the centre of global warming, which in turn is driving climate change, then it is obvious that human or anthropogenic drivers (energy, agriculture, industrial processes, and land-use change/forestry) that are spewing up greenhouse gases are at the centre of the loop [5] while non-anthropogenic drivers are reinforcing the feedback loop [8].

Globally, the need for climate change mitigation strategies, actions, technologies and innovations has never been so urgent than during the current decade. The IPCC AR6 report argues that for most parts of the world and for various representative concentration pathways (RCP 4.5. 8.8), temperatures will continue to rise while precipitation will decline [4], creating important water deficits in many countries. Also, various economic sectors continue to spew out huge amounts of greenhouse gases (GHG). For example, the US Environmental Protection Agency (EPA) [9] presents the following global GHG emission levels for various economic sectors; electricity (25%), agriculture and forestry (24%), transportation (14%), other energy

(10%) and building (6%). From these statistics, at the global scale, agriculture is the second major contributor to the global GHG budget. However, over the last three decades, worldwide GHG emissions from the energy sector has increased four times than agriculture. In fact, though the contribution of various sectors may vary, the role of global greenhouse gas emissions in driving global warming and climate change has been established [8].

It has been agreed that global temperature rise caused by increased greenhouse gases is driving global warming and, consequently, climate change. These changes are impacting agriculture, forests, and water levels at groundwater and river levels, while in other regions, they are shifting the frontiers of established climate zones. In many parts of the world, droughts, floods, and sandstorms are frequent, and livelihoods are negatively impacted [9]. It can be said that while rising temperatures are causing droughts in Africa, they are melting Arctic ice in Canada, and this is negatively impacting other livelihoods that are Arctic ice dependent but also creating more favourable conditions for greenhouse crop production in the Arctic [10], and field crops production in countries. In other parts of the world, climate change has been of immense benefit to agriculture as the growing season of some crops has increased while that of others has been shortened due to prolonged droughts [11]. Increased carbon dioxide concentration, for example, has been argued to increase photosynthesis and therefore yield, particularly C4 plant species. Several studies across the Sahel, for example, have shown that the latter region is getting greener and wetter due to NDVI-based evidence; a change that is partly attributed to increased carbon dioxide emissions and increased photosynthesis [10–12]. Notwithstanding these recent observations, the question that comes to mind is, why is the region still facing acute food security challenges despite these gains in greening and rainfall? This invariably shows that, besides climate change, systems of land use are evidently playing an important role in the Sahel, for example, because, on the one hand, the current recovery in terms of rainfall is not readily available for agriculture due to duration and timing while on the other hand, the rainfall gains are weathered by poorly managed land use systems.

Evidently, most of the reports on the effects of climate change on livelihoods are often from the global south, with Africa being the least responsible for emitting greenhouse gases yet the most vulnerable. This is explained by high levels of vulnerability, which are often anchored on the low adaptive capacity of the people [13, 14]. The adaptive capacity will therefore continue to play an important role in determining how societies will build resilience to climate change. This is seen as the issue at stake now, and in the years to come will no longer be the magnanimity of the climate forcing but, most importantly on, key proxies of adaptive capacity such as poverty and literacy rates, which will evidently determine the ability of a society to access the resources needed to build resilience against such climate shocks [1–3, 15, 16]. Even in the developed north, communities with lower adaptive capacities have been found to be more vulnerable to the effects of climate change due mainly to limited access to resilience-building schemes which are mainly determined by adaptive capacity. The Arctic of Canada, for example, is disproportionately impacted by the effects of climate change mainly due to the extremely low levels of adaptive capacity recorded in such regions when compared to the southern communities in Canada [17].

Therefore, apart from the need to enhance adaptive capacities across most of the south and in the more remote and less developed parts of the world in general that are more vulnerable to climate change, there is equally a strong need for research in these regions or communities. The research will help enhance our understanding of climate change in these regions and therefore help frame how adaptations can be structured.

A major bulwark that is often associated with the south, especially in the context of climate change, is the availability of climate data which is often relevant to better understanding climate change to elaborate robust adaptations. Across Africa, for example, there are inadequate observed weather stations to monitor various climatic variables such as precipitation and temperature. Additionally, in the face of food security, it is not clear how crop yields have evolved over time. This hampers effort at understanding the effects of climate on crop yields and water resources. This phenomenon is more acute in Africa due to inadequate funding to increase the number of weather stations and comprehensive yield accounting.

2. Conclusion

To grapple with these research challenges, scientists can now compare satellite-based precipitation, temperature and crop yield data with existing observed precipitation, temperature and yield data. This process will establish a relationship between satellite-based precipitation, temperature and crop yield with the limited observed data. It becomes feasible to determine which satellite products simulate the limited observed data best. This will yield new knowledge on which satellite product(s) are closest to or best simulate the existing observed data. Such products can now be used as a proxy for precipitation, temperature or yield at any given scale. In the Tensift basin of Morocco for example, among several satellite precipitation products (PERSIANN, PERSIANN CDR, TRMM3B42, ARC2, RFE2, CHIRPS and ERA5), the PERSIANN CDR product is the most representative product as it simulates observed precipitation better than the other products [18]. In agriculture, observed time series yield data from FAOSTAT, and Global Yield Gap Atlas can be compared with crop yield satellite products such as MODIS, Copernicus Sentinel 1 and 2 to determine a proxy for yield. This approach is highly recommended to bridge the wide gap between inadequate observed data and satellite-based data for elements of climate such as precipitation.

Therefore, there is no doubt therefore why, in this book, '*Climate change: Recent Observations,*' the range of chapters varies as does the scope of the book itself, creating room and inclusivity for topics that reflect the great diversity of climate change and its challenges. With a wide range of topics covered, this work will become a key compendium for most enthusiasts of climate change. The major advantage being it covers observations of climate change across the world.

Author details

Terence Epule Epule[1,2]*, Vincent Poirier[1] and Wiam Salih[2]

1 Unité de Recherche et Développement en Agriculture et Agroalimentaire de l'Abitibi-Témiscamingue (URDAAT), Université du Québec en Abitibi-Témiscamingue (UQAT), Notre-Dame-du-Nord, QC, Canada

2 International Water Research Institute, Mohammed 6 Polytechnic University, Ben Guerir, Morocco

*Address all correspondence to: terenceepule.epule@uqat.ca

References

[1] Doukas H, Patlitzianas KD, Kagiannas AG, Psarras J. Renewable energy sources and rationale use of energy development in the countries of GCC: Myth or reality? Renewable Energy. 2006;**31**(6):755-770

[2] Maslin M. Climate Change: A Very Short Introduction. Oxford, UK: OUP Oxford; 2014

[3] Price M. Mountains: A Very Short Introduction, 2015. Oxford, UK: OUP Oxford; 2015

[4] Intergovernmental Panel on Climate Change (IPCC). The Physical Science Basis Regional fact sheet - Africa. Sixth Assessment Report of Working Group 1-The Physical Science Basis. Sith Assessment Report of Working Group I. Available online at: https://www.ipcc.ch/report/ar6/wg1/ [Accessed: 18 April 2023]

[5] Stern DI, Kaufmann RK. Anthropogenic and natural causes of climate change. Climatic Change. 2014;**122**:257-269

[6] Crowley TJ. Causes of climate change over the past 1000 years. Science. 2000;**289**(5477):270-277

[7] Davenport FV, Diffenbaugh NS. Using machine learning to analyze physical causes of climate change: A case study of US Midwest extreme precipitation. Geophysical Research Letters. 2021;**48**(15):e2021GL093787

[8] Fakana ST. Causes of climate change. Global Journal of Science Frontier Research: H Environment & Earth Science. 2020;**20**(2):7-12

[9] United Sates Environmental Protection Agency (USA EPA). Greenhouse Gas Emissions and Removals. Available online at: https://www.epa.gov/ghgemissions [Accessed: 15 April 2023]

[10] Herrmann SM, Anyamba A, Tucker CJ. Recent trends in vegetation dynamics in the African Sahel and their relationship to climate. Global Environmental Change. 2005;**15**(4):394-404

[11] Leroux L, Bégué A, Seen DL, Jolivot A, Kayitakire F. Driving forces of recent vegetation changes in the Sahel: Lessons learned from regional and local level analyses. Remote Sensing of Environment. 2017;**191**:38-54

[12] Epule ET, Peng C, Lepage L, Chen Z. The causes, effects, and challenges of Sahelian droughts: A critical review. Regional Environmental Change. 2014;**14**:145-156

[13] Rao N, Lawson ET, Raditloaneng WN, Solomon D, Angula MN. Gendered vulnerabilities to climate change: Insights from the semi-arid regions of Africa and Asia. Climate and Development. 2019;**11**(1):14-26

[14] Rhoné B, Defrance D, Berthouly-Salazar C, Mariac C, Cubry P, Couderc M, et al. Pearl millet genomic vulnerability to climate change in West Africa highlights the need for regional collaboration. Nature Communications. 2020;**11**(1):5274

[15] Epule TE, Chehbouni A, Dhiba D, Moto MW, Peng C. African climate change policy performance index. Environmental and Sustainability Indicators. 2021a;**12**:100163

[16] Epule TE, Chehbouni A, Dhiba D, Moto MW. The readiness index for

climate change adaptation in Africa: The role of climate and adaptive capacity proxies. Applied Sciences. 2021b;**11**(20):9413

[17] Ford JD, Smit B, Wandel J. Vulnerability to climate change in the Arctic: A case study from Arctic Bay, Canada. Global Environmental Change. 2006;**16**(2):145-160

[18] Salih W, Chehbouni A, Epule TE. Evaluation of the performance of multi-source satellite products in simulating observed precipitation over the Tensift Basin in Morocco. Remote Sensing. 2022;**14**(5):1171

Section 2

Climate Risks, Resilience and Vulnerability

Chapter 2

Spatiotemporal Variation Analysis of Arctic Sea Ice Thickness and Volume Based on CryoSat-2 Altimetry Satellite Data

Xubing Zhang, Ruizhi Li, Wenxi Li, Liang Zhang and Kai Wang

Abstract

Sea ice monitoring is helpful to the research of the Arctic channels and climate environment. Through echo signal classification, re-tracking correction, and other techniques, data from the Cryosat-2 radar altimetry satellite between 2010 and 2020 facilitated the retrieval of Arctic sea-ice thickness and volume. The study subsequently analyzed the variations in the average thickness and volume of first-year ice and multi-year ice in the Arctic sea ice over the past decade. In the past decade, the volume of sea ice in the Arctic region fluctuates slightly. The multiyear ice changed greatly in 2013, while the first-year ice shows an increased rate of both winter growth and summer ablation. The presence of uncertainties arising from the fluctuations in sea ice density values and the intricate patterns of Arctic circulation might impart subtle biases in the measurements of sea ice thickness. However, the sea ice freeboard and thickness data inverted based on the Cryosat-2 data were validated by the data obtained through the Operation Ice Bridge (OIB), and the validation results indicated that they are correspondent.

Keywords: sea ice, Cryosat-2, the Arctic, thickness, volume

1. Introduction

As an indicator of the Arctic climate, the sea ice is affected by the climate change and reacts to global climate. It is an important indicator of global climate change research and a window to understanding the current situation and change trend of the Arctic environment. For a long time, limited by the harsh natural environment of the Arctic region, the Arctic sea ice observation is high cost and low efficiency. Therefore, it is difficult to realize the large-scale, multi-scale, and long-time series monitoring. With the rapid development of remote sensing technology, satellite altimetry has become an effective method to continuously obtain the accurate global thickness information on the sea ice [1]. The Earth Explorer Opportunity Mission-2 (CryoSat-2) satellite which launched in April 2010, with full coverage of the Arctic region and a

 IntechOpen

centimeter-level precision [2], could meet the need for long-term monitoring of the sea ice thickness.

At the same time, the theoretical method of the altimetry satellite inversion of the sea ice thickness has also attracted the attention of many researchers. In 2003, the sea ice thickness in the 40°N ~ 81.5°N range was retrieved by Peacock, Laxon et al. based on the Earth Resources Satellite-1 altimetry satellite [3, 4]. As mentioned in the articles [3, 4], the sea ice was distinguished from the water according to the pulse peaks and backscattering coefficients of the radar echo signals. After that, the echo signals are corrected by the re-tracking algorithm to calculate the elevation of the sub-satellite point and obtain the elevation difference between the sea ice and the surrounding seawater namely sea ice freeboard. Then, the sea ice thickness is estimated based on the static equilibrium state of the sea ice, and some parameters such as snow density and snow depth as well. Currently, the echo signals classification [5–8], re-tracking correction algorithm [9–12], and sea ice thickness inversion model parameters [13–16] have become hot issues in sea ice thickness monitoring.

To study the changes in the Arctic sea ice thickness between 2010 and 2019, this paper selects the following data such as the CryoSat-2 data with high accuracy and full coverage, the MASAM2 sea ice density product with daily update of the 4 km resolution grid, the OSI SAF (Ocean and Sea Ice Satellite Application Facility) daily update product of the sea ice type with 10 km resolution grids, the UCL13 MSS (Mean Sea Surface model of the University College London 2013) and the ice snow thickness data with the 24 km resolution grids of daily update produced by the Canadian Meteorological Center. With the above data, this paper analyzes the variation characteristics of the sea ice thickness in recent years, which could provide the data basis for the change trend research of the large-scale sea ice thickness in the Arctic region.

2. Sea ice thickness inversion data

2.1 Cryosat-2 altimetry satellite data

CryoSat-2 was launched by the ESA on April 8, 2010. The Synthetic Aperture Radar Altimeter (SIRAL), carried by CryoSat-2, includes three modes of operation: Low-Resolution Measurement Mode (LRM), Synthetic Aperture Radar Mode (SAR), and Synthetic Aperture Radar Interferometry Mode (SARIn). The LRM model is primarily used to make measurements of continental ice caps and the vast majority of the Earth's ice-free ocean and land areas. The SAR model measures mainly the sea ice and some ocean basins and coastal areas. The SARIn model measures mainly the ice cap, mountain glaciers, ground-transfer currents, hydrographic basins, and the coastal zone regions. In this paper, the L1B data of the SAR model and SARIn model are used to estimate the Arctic Sea ice thickness before September 2014, and the L1B data of the SAR model is used to retrieve the whole Arctic Sea ice thickness after this time point.

2.2 Sea ice density data

The sea ice density product is used to classify the CryoSat-2 radar satellite echo signal waveforms. We use the SSM/I_SSMIS dataset provided by the National Snow and Ice Data Center (NSIDC) to obtain sea ice density data from October 2010 to March 2012, which provides daily and monthly updates of 25 km gridded sea ice density data [17, 18]. After October 2012, the MASAM2 (MASIE-AMSR2) dataset provided by NSIDC was used. Compared with the AMSR2, the MASAM2 data are less affected by the weather, have higher accuracy and resolution, and are more accurate in calculating the sea ice density during sea ice melt. The visualization of the MASAM2 data is shown in **Figure 1**.

2.3 Sea ice type data

The sea ice type data could assist in calculating the sea ice thickness. We use the Ocean and Sea Ice (OSI SAF) dataset produced by the Norwegian Meteorological Service's Ocean and Sea Ice Satellite Application Facility (OSSAF), which is a daily updated 10 km grid sea ice type product from 2005 to the present [19]. A schematic diagram of the sea ice type data visualized from the OSI SAF for March and October 2018 is shown in **Figure 2**.

2.4 Sea ice snow thickness data

As the input parameter of the sea ice static equilibrium model, the sea ice snow thickness data are directly related to the accuracy of the sea ice thickness calculation

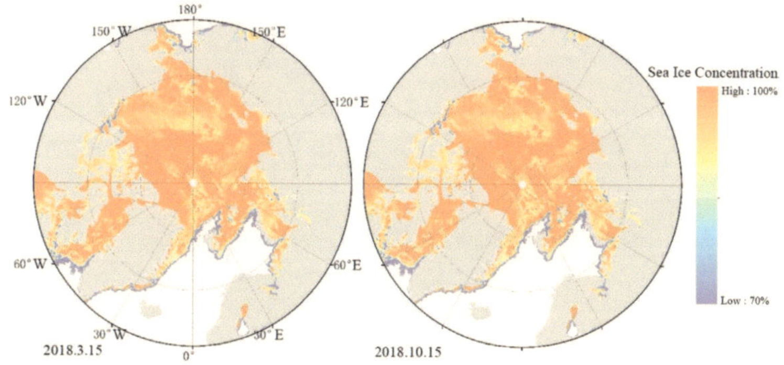

Figure 1.
The MASAM2 sea ice density data.

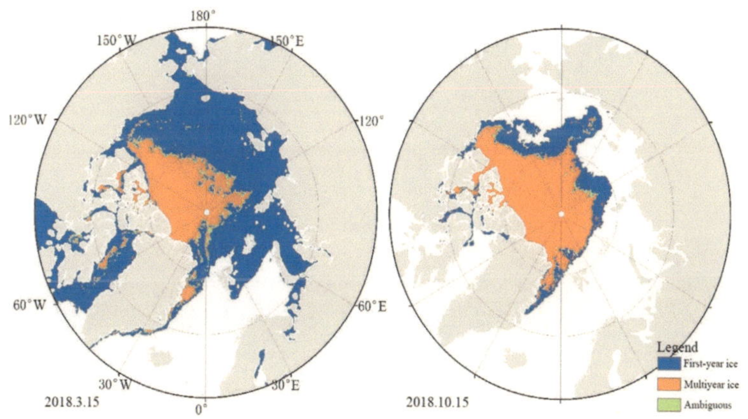

Figure 2.
Schematic diagram of the OSI SAF sea ice type products in March and October 2018.

results. The daily updated snow thickness data from the Canada Meteorological Centre (CMC), is a dataset based on the snow depth data provided by the World Meteorological Organization Information System (WMOIS) and daily snow thickness measurements in the field [20]. A schematic of the monthly average snow thickness data for March and October 2018 for this dataset is shown in **Figure 3**.

2.5 The operation ice bridge data

The Operation IceBridge (OIB) science data are used to validate the results of the sea ice thickness inversion process. IceBridge L4 sea ice freeboard height, snow depth, and ice thickness data (IDCSI4, IceBridge L4 Sea Ice Freeboard, Snow Depth, and Ice Thickness) are used to validate the sea ice freeboard and thickness inversion results of this study [21, 22]. The March 2013 OIB flight path is shown in **Figure 4**.

2.6 Arctic mean sea surface data

The Mean Sea Surface in the Arctic region is one of the parameters used for the sea ice freeboard estimation. We use the UCL13 mean sea surface model produced by

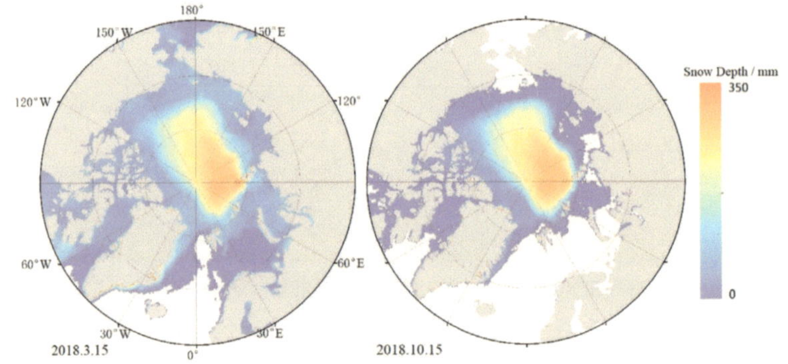

Figure 3.
Schematic of the Canadian Meteorological Centre snow thickness products in March and October 2018.

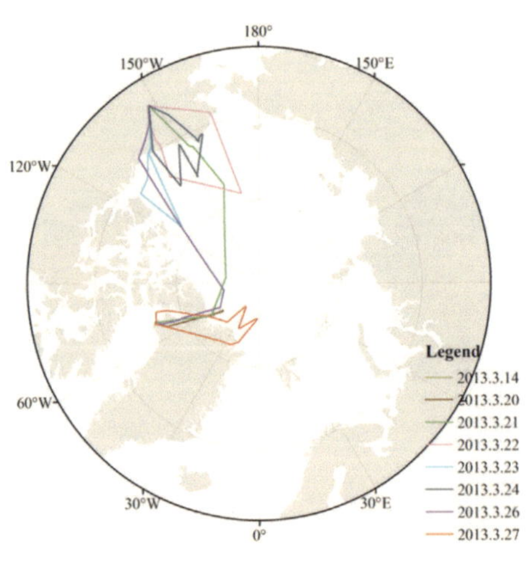

Figure 4.
Schematic diagram of the March 2013 OIB flight path.

University College London (UK) in 2013. Because previous mean sea surface models of the Arctic lacked satellite data north of 81.5°N, the UCL13 is more accurate over the entire Arctic region [23–25].

3. Sea ice thickness inversion

The thickness and volume of the sea ice in the Arctic region are estimated in this paper by using L1b level data from the Cryosat-2 radar altimetry satellite, and the main steps are illustrated in **Figure 5**. First, the echo signal data in the Arctic sea region are denoised. Then, the echo signal classification and re-tracking correction are performed, and the sub-satellite point elevation is calculated. Finally, the sea ice thickness is calculated based on the sea ice static equilibrium model, supplemented by the snow depth and the sea ice density data.

3.1 Data preprocessing

The satellite altimeter transmits and receives the reflected pulse signals to the sub-satellite point, and the distance from the satellite altimeter to the sub-satellite point is obtained by calculating the time difference between the transmitted and received pulse signals [26–28]. Where the echo signal waveform reflected from the sub-satellite point is recorded in the CryoSat-2 satellite L1b data, the shape of which is shown in **Figure 6**. The echo signal information needs to be pre-processed to reduce the background radiation, the instrument noise, or the waveform anomalies. The satellite data in the region north of 60°N latitude is also screened to remove waveforms with missing or mutilated information.

3.2 Echo signal classification

The pulse-peakiness (PP) and the pulse standard deviation so-called stack standard deviation (SSD) are used to distinguish between the sea ice and the seawater in this paper [6]. The pulse peak value is used to describe the shape of the echo signal, and

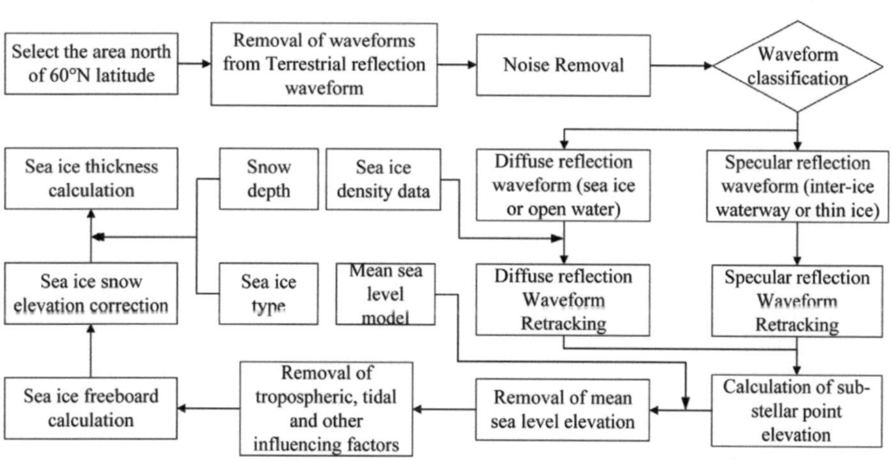

Figure 5.
Inversion process of the sea ice thickness.

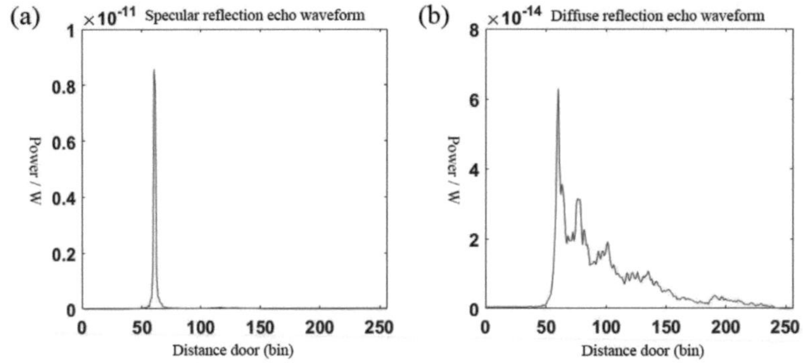

Figure 6.
Cryosat-2 measurement waveform. (a) is the specular echo; (b) is the diffuse echo.

the larger the value is, the peak exists in the waveform. The pulse standard deviation is the standard deviation of the scattering direction of the echo signal. And the larger the value is, the more dispersed the echo signal reflected from the surface is, conversely, the reflection direction of the echo signal is concentrated. Pulse standard deviation values are provided in the L1b file, and the *PP* value is calculated using the following equation:

$$PP = \frac{P_{max}}{P_{mean}} \quad (1)$$

Where *PP* is the pulse-peakiness, P_{max} is the maximum power in the echo signal, and P_{mean} is the average power in the echo signal. Specular echoes occur when the radar pulses are reflected from a smooth, mirror-like surface such as the sea ice leads or very thin ice (**Figure 6a**). Diffuse echoes occur when the radar pluses are reflected from a rougher surface such as an ice floe or the open ocean (**Figure 6b**).

Specular echoes are defined as those with a pulse-peakiness greater than 18 and the SSD less than 6.29 for SAR mode echoes or the SSD less than 4.62 for SARIn mode echoes. Diffuse echoes are those with a pulse-peakiness less than 9 and the SSD greater than 6.29 for SAR mode echoes or the SSD greater than 4.62 for SARIn mode echoes [6]. Subsequently, it is necessary to distinguish the echoes of the sea ice from the echoes of the ocean, as both are classified as diffuse reflection waveforms. Additionally, the sea ice concentration products MASIE-AMSR2 used in this paper only recorded the regions with sea ice concentration above 70%. We thereby define the sea ice regions as those with a sea ice concentration greater than 70%, or else, the diffuse echoes are not trusted to come from either sea ice or open ocean and are removed from the processing.

3.3 Echo signal re-tracking correction

The distance between the satellite altimeter and the sub-satellite point is calculated by the time interval between the satellite altimeter transmitting pulse signal and receiving the echo signal reflected by the ground object. Generally, the time tracking point of the transmitting signal has been set on the satellite. Similarly, the time tracking point of the echo signal is pre-set at the midpoint of the waveform sampling window. However, the actual tracking point would always be offset from the original set tracking point, but the round-trip time of the pulse signals is still calculated according to the positions of the original set tracking points. That it is necessary to correct the distance differences between the actual tracking points and the preset tracking points to obtain high-precision height measurement values [29]. In this paper, a threshold first maximum re-tracker (TFMRA) is adopted, which is based on the empirical estimation of statistical [15]. The flow chart is shown in **Figure 7**. The main formulas are shown in Eqs. (2)–(5):

$$P_{noise} = \frac{1}{noise}\sum_{i}^{noise} P_i \tag{2}$$

$$\begin{aligned} &A = \max\left((P_n)_{First}\right), n = 1, 2, 3, \dots, N \\ &(A > 0.15 + P_{noise}) \end{aligned} \tag{3}$$

$$T_h = P_{noise} + q \times (A - P_{noise}) \tag{4}$$

$$Re = G_{k-1} + \frac{T_h - P_{k-1}}{P_k - P_{k-1}} \tag{5}$$

where *noise* is the number of the range bins, and the value in the Cryosat-2 echo signal is generally 6 P_{noise} is the noise value of echo waveform. The average value of the echo power corresponding to the first six range bins is taken as the noise value P_{noise}. P_i is the echo power corresponding to the range bin i. A is the first peak of the echo waveform, and it needs to be larger than the sum of 0.15 and the value of P_{noise}. P_n is the power value of the normalized echo waveform. n is the number of the range bins, and the total number of the range bins is 256 in the SAR mode, and 1024 in the SARIn mode. q is the proportional coefficient namely the threshold value. T_h is the power value of the re-tracking point. Re is the range bin number obtained after re-tracking

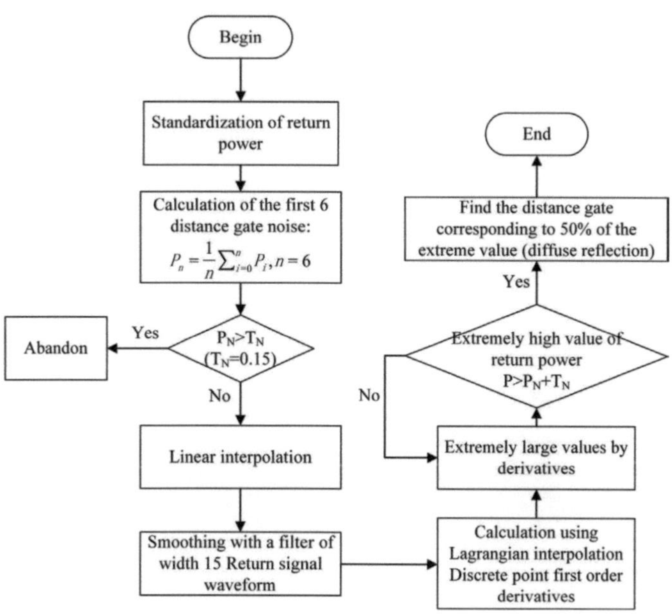

Figure 7.
Flow chart of TFMRA.

correction, and if it is between two range bins, it would be calculated by linear interpolation. G_{k-1} is the previous range bin corresponding to *Re*.

3.4 Sea ice freeboard inversion and verification

After re-tracking correction, the sea ice and seawater elevation could be calculated, and then the sea ice freeboard value is obtained. **Figure 8** shows the schematic diagram. The orbit elevation h_{cs} of the altimetry satellite is known and stored in the L1b file. First, the distance *H* between the sub-satellite point and the altimetry satellite is calculated by the round-trip time of the radar pulse. The difference between h_{cs} and *H* is the elevation value of the sub-satellite point under the reference of the WGS84 ellipsoid. Then the sea ice freeboard is calculated according to the elevation difference between the sea ice and the sea ice leads. Although some experiments have shown that the scattering layer of the radar pulse at the ice-snow interface of the sea ice with snow cover varies with the increase in temperature [30]. However, the variation of the scattering layer has little effect on the results in the annual sea ice observation [31]. Therefore, we assume that the radar pulses are scattered at the ice-snow interface, and the formulas are shown in Eqs. (6)–(8):

$$h_f = h_{CS} - h_{ssh} - H - C_G \tag{6}$$

$$H = \frac{ct_n}{2} + C_R \tag{7}$$

$$C_R = (b_O - b_n)\Delta_b \tag{8}$$

Where h_f is the freeboard of the sea ice, representing the height of sea ice above sea level. H_{ssh} is the elevation of the local sea level. And this paper refers to the elevation value of the Arctic sea surface model UCL13. C_G is the influence of various factors in the atmosphere on a radar range, such as dry troposphere, wet troposphere, ionosphere, ocean tide, geocentric polar tide, and so on. *C* is the speed of light. t_n is the time in which the radar pulse is transmitted from the satellite and reflected by the surface of the sub-satellite point, and then received by the satellite. C_R is the re-tracking correction value. b_O is the number of the range bin located in the middle,

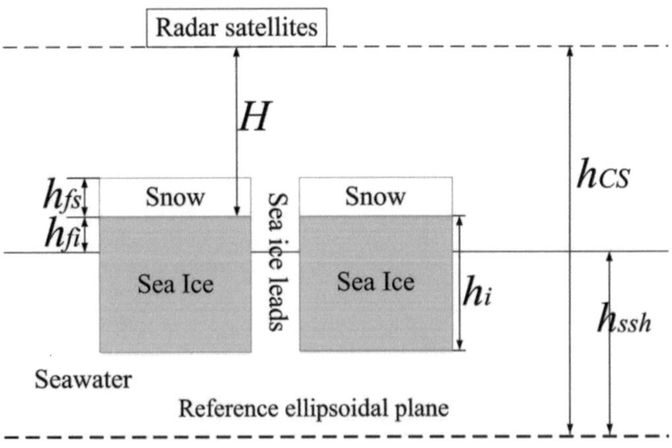

Figure 8.
Schematic diagram of sea ice altimetry.

which is 128 in SAR mode and 512 in SARIn mode. b_n is the number of the bin determined by waveform re-tracking and Δ_b is the bin width (0.2342 m).

Because the electromagnetic waves will travel slower when they move into the snow layer, it will cause errors in calculating the freeboard. It is necessary to modify the calculation of the sea ice freeboard. The corrected sea ice freeboard is shown in Eq. (9) [31]:

$$f_c = f_i + h_s\left(\frac{c_v}{c_s} - 1\right) \tag{9}$$

where f_c is the corrected sea ice freeboard. f_i is the original sea ice freeboard. h_s is the snow depth. c_v is the speed of light propagation in a vacuum (3 × 108 m/s). And c_s is the speed of light propagation in the snow (2.4 × 108 m/s). Therefore, the corrected freeboard is simply expressed as:

$$f_c = f_i + 0.25 \times h_s \tag{10}$$

To comprehensively evaluate the effect of the echo signal correction based on TFMRA, the sea ice freeboard values, which were inverted by TFMRA based on the different threshold q Eq. (4) of the sea ice, are compared with that of the airborne OIB data. The verification time selected is March 2013, during which the flight missions of the airborne OIB data are only carried out within seven days from March 20 to 24, and 26 to 27. The flight route was in the sea area of northern Canada. Cryosat-2 satellite data from March 18 to 29 are selected for inversion verification. The horizontal spatial resolution of Cryosat-2 is about 300 meters, while that of OIB data is close to 1 meter, and then the Cryosat-2 data points are sparse and the OIB data points are dense. To compare the two data, a 5 km resolution grid is made based on the OIB data. The average value of data points in the grid is used to represent the value of the grid. The grid containing both Cryosat-2 and OIB data points was selected for comparison. A total of 1172 grids meet the requirements. **Figure 9** shows the freeboard diagram of the sea ice measured by airborne OIB.

The statistics of the survey points in the 1172 grids are shown in **Table 1**. The number of the altimetry points of the Cryosat-2 in the grid is far less than that of the

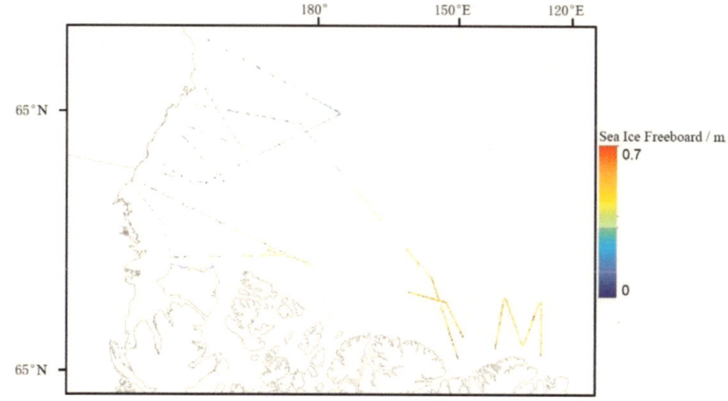

Figure 9.
Schematic diagram of sea ice freeboard measured by airborne OIB in March 2013 (Using WGS 1984 North Pole LAEA Atlantic projection coordinate system).

	Total number	Mean value	Maximum value	Minimum	Median
OIB high points	78,696	67.20	155	1	69
Cryosat-2 high points	12,034	10.28	42	1	8

Table 1.
Inversion of the number of high points in 1172 grids.

airborne OIB, and the altimetry points of both data are not evenly distributed in the grids. Inevitably, there are some errors between the measured values of the Cryosat-2 data and that of the airborne OIB. Comparing the altimetry data of Cryosat-2 within 10 days and the measured data of the OIB within 7 days. Because of the inconsistent timing of the acquisition of the two types of data, it is possible that the sea ice thickness changed during this period, resulting in some errors between the Cryosat-2 and the OIB data.

Calculating the sea ice freeboard requires the local sea level elevation. We use UCL13 mean sea level elevation model produced by the University of London to obtain the sea level elevation of sea ice. The threshold q Eq. (4) is significant to the results of the TFMRA. Therefore, the threshold q of the sea ice was set to 0.5, 0.6, 0.7, and 0.8 to correct the sea ice echo signal by TFMRA, respectively. The sea ice freeboard values are calculated based on the UCL13 sea level average elevation model and the Canadian snow data set, and the results compared with the OIB measurements are shown in **Table 2**.

The scatter plots based on the airborne OIB measurements and the inversion freeboard values on different thresholds are shown in **Figure 10**. Every scatter point denotes the inversion freeboard mean value and the OIB measurement mean value of the altimetry points in the same grid abovementioned. The scatter plot shows the difference in the number and distribution of the inversion freeboard values and OIB measurements in the grids. Ideally, the more points are concentrated in the line of y = x, the closer the inversion freeboard values are to the OIB measurements, which is more consistent with the actual situation.

As shown in **Figure 10a**, the scatter points are concentrated in the lower right area, which means that the inversion freeboard values are generally larger than the OIB measurements, and the largest deviation is 18.7 cm, when the threshold is set to 0.5. As shown in **Figure 10c**, the scatter points are relatively evenly distributed on both sides of the y = x line, which means that the inversion freeboard values are generally

	OIB data/m	TFMRA-50% threshold value/m	TFMRA-60% threshold value/m	TFMRA-70% threshold value/m	TFMRA-80% threshold value/m
Mean value	0.380	0.567	0.463	0.361	0.263
Mean difference	/	0.187	0.083	−0.019	−0.117
RMSE	/	0.257	0.198	0.187	0.222
Correlation coefficient	/	0.512	0.463	0.427	0.385

The significance of "/" is that values are missing.

Table 2.
Comparison of airborne OIB freeboards and sea ice freeboards based on different threshold values of the sea ice according to TFMRA.

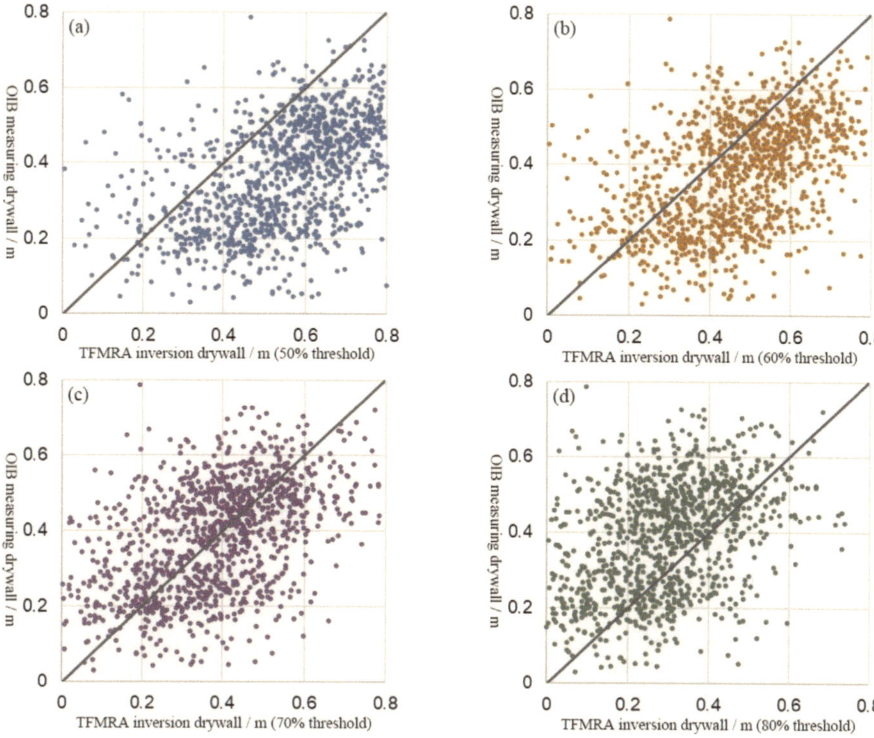

Figure 10.
Freeboard scatters point inversion based on different thresholds of TFMRA and OIB. (a) 50% threshold inversion; (b) 60% threshold inversion; (c) 70% threshold inversion; (d) 80% threshold inversion.

closer to the OIB measurements, and the average deviation is 1.9 cm, when the threshold is set to 0.7. Overall, compared the four plots of **Figure 10** the freeboard inversion results are the most consistent with the actual situation when the threshold is set to 0.7.

3.5 Sea ice thickness inversion and verification

According to the actual situation of the sea ice in the Arctic ocean, it is assumed that the sea ice covered with snow is in equilibrium with the seawater. The sea ice thickness could be calculated by the static equilibrium model [3], as shown in Eq. (11):

$$h_i = \frac{f_c \beta_w + h_s \rho_s}{\rho_w - \rho_i} \tag{11}$$

where h_i is the sea ice thickness, f_c is the sea ice freeboard, and ρ_w is the density of the seawater. In this paper, the seawater density is 1023.9 kg/m^3. h_s is the snow thickness on the sea ice, referring to the snow thickness data of the Canada Meteorological Center. ρ_s is the snow density on the sea ice, referring to Spreen [32], being 330 kg/m^3. ρ_i is the sea ice density, which is 882.0 kg/m^3 for multi-year ice density and 916.7 kg/m^3 for first-year ice density, referring to Alexandrov [33].

In most cases, the hydrostatic equilibrium model is applicable to the calculation of sea ice thickness. However, sometimes, as the thinner sea ice is pressed by the thick

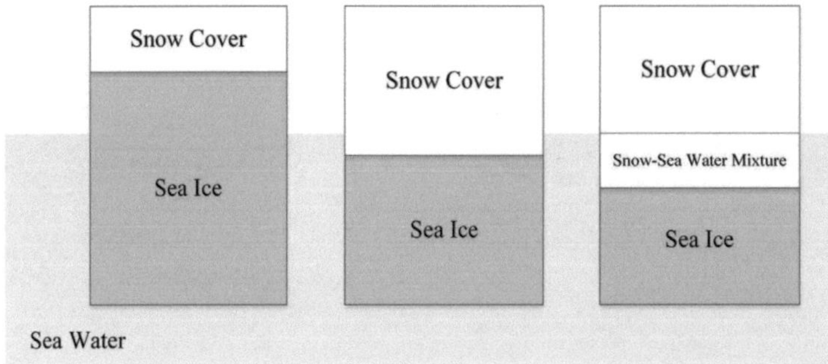

Figure 11.
Static equilibrium model in various cases (from left to right, the freeboard value is positive, zero, and negative).

snow and fluctuates up and down by the wave influence, and the seawater may penetrate the sea ice snow and form a snow-water mixture (**Figure 11**) [34, 35]. Under these circumstances, the sea ice freeboard values measured by the satellite altimeter may be zero or negative.

When the freeboard of sea ice is zero or negative, the properties of the snow-water mixture layer are very different from those of the sea ice and the snow cover. Therefore, when the snow cover, snow water mixture, and the sea ice are in equilibrium in the seawater, the calculation formula of the sea ice thickness is shown in Eqs. (12) and (13) [36]:

$$h_i = \frac{\rho_{slh} - \rho_w}{\rho_w - \rho_i} \times h_{slh} + \frac{\rho_s}{\rho_w - \rho_i} \times h_s \tag{12}$$

$$h_{slh} = -h_f \tag{13}$$

where h_i, h_s, ρ_w, ρ_s and ρ_i are the same as the Eq. (11). h_f is the freeboard height of sea ice, h_{slh} is the thickness of the snow water mixture, and ρ_{slh} is the density of the snow water mixture. The thickness of the snow-water mixture layer is equal to the absolute value of the sea ice freeboard, and its density is 940 kg/m^3 [37].

To verify the availability of the inversion results of the sea ice thickness, the inversion results were compared with the airborne OIB data. The abovementioned 5 km resolution grids based on OIB data are adopted here. The grids containing Cryosat-2 and OIB data points were selected for verification. A total of 1025 grids met the requirements. The sea ice thicknesses measured by the airborne OIB are shown in **Figure 12**. The Canadian snow data product and UCL04 snow thickness model are used to calculate the sea ice thickness, and the results are compared with the airborne OIB measurement data, respectively as shown in **Table 3**.

As shown in **Figure 13,** at sea ice thickness of less than 1 m, there is a large difference between the calculated sea ice thicknesses of the Canadian snow data product and the UCL04 snow thickness model. The points in **Figure 13a** are scattered, while the points in **Figure 13b** are closer to the x = 1 line. Can be introduced on sea ice thinner than 1 meter, UCL04 have larger thicknesses of the snow on the sea ice than that of the Canadian snow data.

When the reversion values of the sea ice thicknesses are greater than 1 m, the points in **Figure 13a** and **b** are distributed consistently and concentrated in the lower right region of the y = x straight line, indicating that the Canadian snow data product and the UCL04 have almost the consistent snow thicknesses.

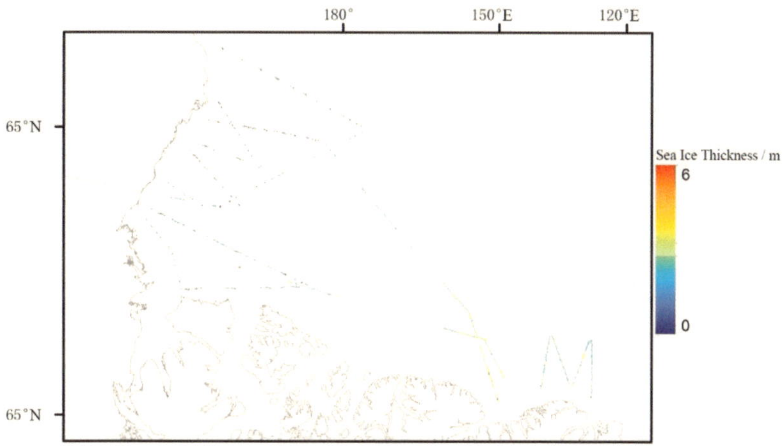

Figure 12.
The sea ice thicknesses were measured by airborne OIB in March 2013 (Using WGS 1984 North Pole LAEA Atlantic projection coordinate system).

	Thickness measured by OIB /m	Sea ice thickness calculated by combining Canadian snow cover thickness/m	Sea ice thickness calculated by combining UCL04 snow cover thickness/m
Mean value	2.239	2.764	2.961
Mean difference	/	0.525	0.722
RMSE	/	0.882	1.085
Correlation coefficient	/	0.756	0.675

The significance of "/" is that values are missing.

Table 3.
Comparison of the inversion values of the sea ice thicknesses based on different snow products and the sea ice thickness measured by airborne OIB.

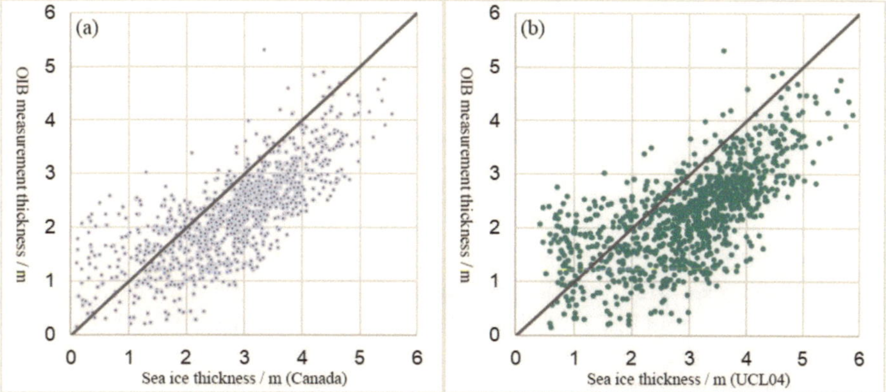

Figure 13.
Scatter plot of inversion values of the sea ice thicknesses based on different snow products and the sea ice thicknesses measured by OIB.

Combined with **Table 3**, the average reversion values of the sea ice thicknesses based on the Canadian snow data are closer and more correlated to the airborne OIB measurements. The slightly larger reversion values than the OIB measurements may be due to the low signal-to-noise ratio of the snow radar echo signal in the ice ridge region used in the OIB, which resulted in the underestimation of the actual sea ice thickness by the airborne OIB [38]. In summary, the Canadian snow data products are largely reflective of Arctic Sea ice thickness variability.

In summary, compared to the sea ice inversion values based on the UCL04, the inversion values based on the Canadian snow data have a smaller average difference and a higher correlation coefficient with the OIB measurements, therefore the Canadian snow data product was adopted here to invert the sea ice thickness.

4. Sea ice thickness and volume change analysis in the arctic

4.1 Analysis of Arctic Sea ice thickness variation

According to the above steps, the Arctic sea ice thickness data are inverted based on the Cryosat-2 data. It is notable that since the limitation of the sea ice concentration products MASIE-AMSR2 (Section 3.2) the inversion sea ice thickness data are distributed only in the regions where the sea ice concentration is greater than 70%. And then the inversion sea ice thickness data are divided using a 5 km resolution grid north of 60°N (Stereographic North Pole for projection). The sea ice thickness of each grid is represented by the average of the sea ice thickness of the data points within the grid, and if there are less than 5 available altimetry points in this grid, the average of the sea ice thicknesses of the surrounding 8 grids would be taken as the sea ice thickness of this grid. Studies show that the annual sea ice coverage area was greatest in March, and as temperatures rose in April, the sea ice continued to melt and the sea ice coverage area and thickness decreased to the lowest point in October [39]. Therefore, to highlight the change of sea ice, the sea ice thicknesses in March and October are inverted and compared in this paper.

Figures 14 and **15** depict the schematic distribution of the sea ice thickness in October and March in the last decade. The sea ice coverage is almost all in the core area of the Arctic Ocean. In October, there is almost no sea ice grid in the Chukchi, Beaufort, and the Barents Seas, and only a few in the East Siberian and the Kara Seas. **Figure 16** depicts the grids of the different sea ice thicknesses, where the horizontal coordinate is the thickness of the sea ice, the vertical coordinate is the number of grids, and the different colored polygonal lines represent different years.

We can see that, in October, the sea ice is mostly 1–2 meters thick, while in March, it is around 2 meters thick.

According to **Figure 16a**, sea ice in October 2016 exceeded subsequent years in terms of the grid amount and the thickness, which showed a decreasing trend after 2016. In particular, the grid amount of the sea ice thickness being no less than 2 m was the lowest in October 2011, and the grid amount of the sea ice thickness is less than 2 m was lowest in October 2018. **Figure 16b** shows that the grid amount of the sea ice thickness arrives to the highest peak when the sea ice thicknesses are around 2 m in March of each year. Furthermore, the highest peak grid amount was the largest in March 2020 and lowest in March 2018.

The average sea ice thicknesses in March and October from 2010 to 2020 are shown in **Figure 17**. There are large fluctuations in the average sea ice thickness [40].

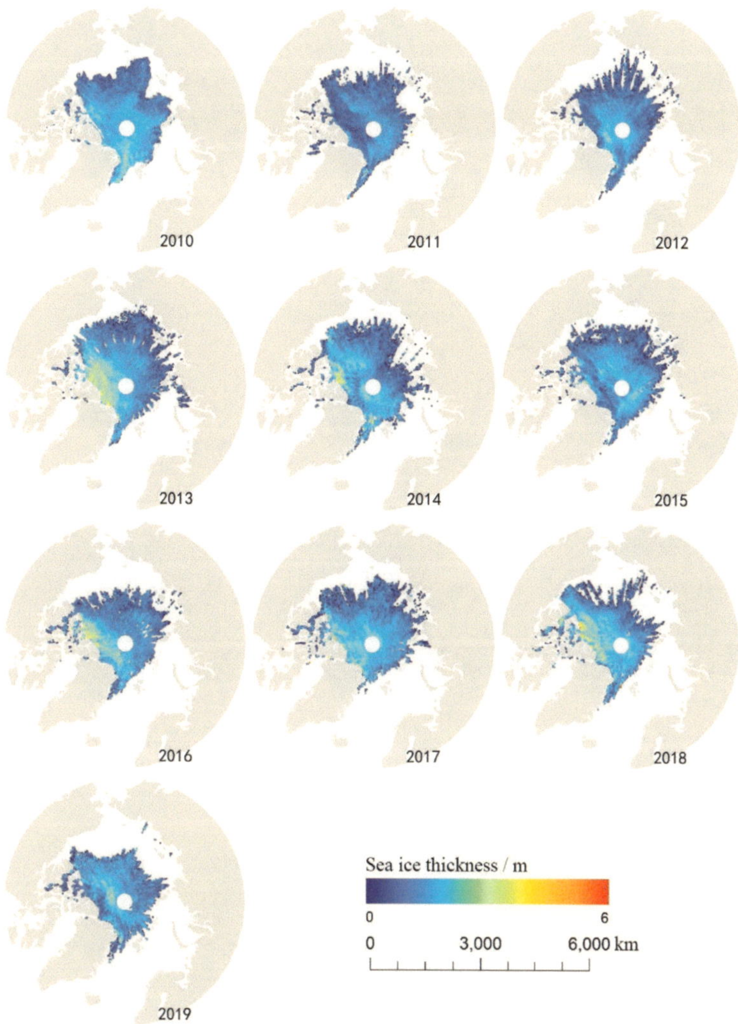

Figure 14.
Sea ice thickness distribution in October from 2010 to 2019.

In October 2011, compared to October 2010 the average sea ice thickness decreased by 26.13%, or 0.38 m, which is also the largest decrease in the average sea ice thickness in nine years. In October 2013, compared to October 2012 the average sea ice thickness showed a significant increase of 22.88% or 0.27 meters. In March 2014, compared to March 2013 the average sea ice thickness increased by 15.55%, or 0.29 m. It reached the maximum value of the average sea ice thickness in the past nine years. In March and October 2015, compared to October 2014 the average sea ice thickness decreased by 2.29% and 7.75%, respectively. In October 2016, the average sea ice thickness showed a significant increase of 12.57%. And in March and October 2017, the average sea ice thickness changed very slightly, with the changes both within 1%. In October 2018, the average sea ice thickness showed a small increase with 10.71% growth. In March 2019, the average sea ice thickness increased by 5.4%. In March 2020, the average sea ice thickness decreased by 7.82%. The average sea ice thickness showed a significant increase in March 2014 and a small decreasing trend in the rest of the

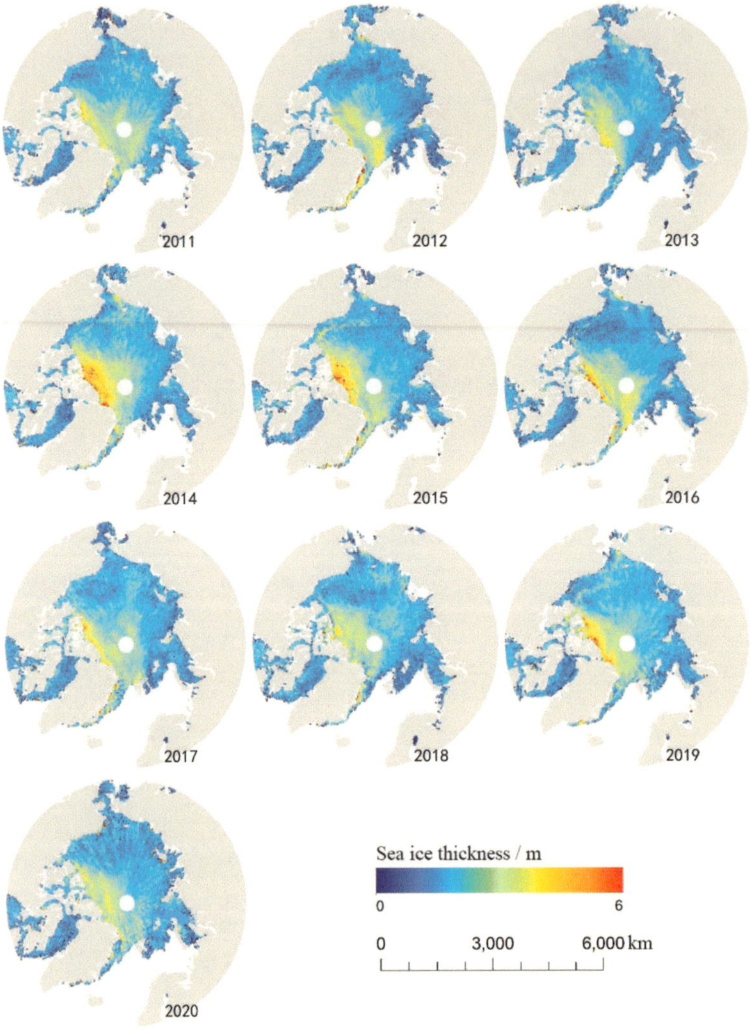

Figure 15.
Sea ice thickness distribution in March from 2011 to 2020.

years, with an average annual change rate of −0.51%. Overall, the variation of the average sea ice thickness in March is relatively stable over the years.

October is the end of autumn in the Arctic region, and the Arctic will enter a long winter in November. After the melting of the peripheral sea ice in summer, only the sea ice remains in the core of the Arctic region. The average sea ice thickness change in October is consistent with the trend of the multiyear ice volume change. Before 2013, the average sea ice thickness was highly variable. After 2013, the average sea ice thickness decreased slightly in 2015 and increased slightly in 2018. The annual mean sea ice thickness decreased by 0.012 m in October 2019 compared to October 2010, with an average change of 0.18%. In general, there are small fluctuations in the average thickness change of sea ice in October over the years, and the overall trend is stable. This is consistent with the findings of Hiroshi Sumata et al. [41] and Feng XIAO et al. [42] on Arctic sea ice thickness.

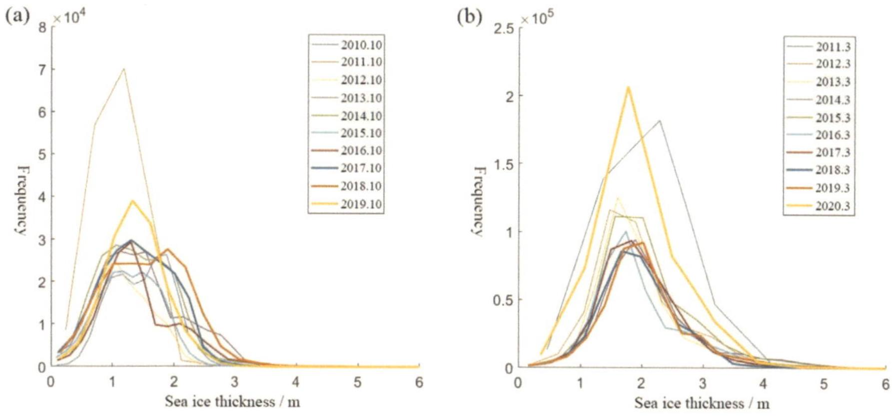

Figure 16.
Frequency and sea ice thickness in March and October from 2010 to 2020.

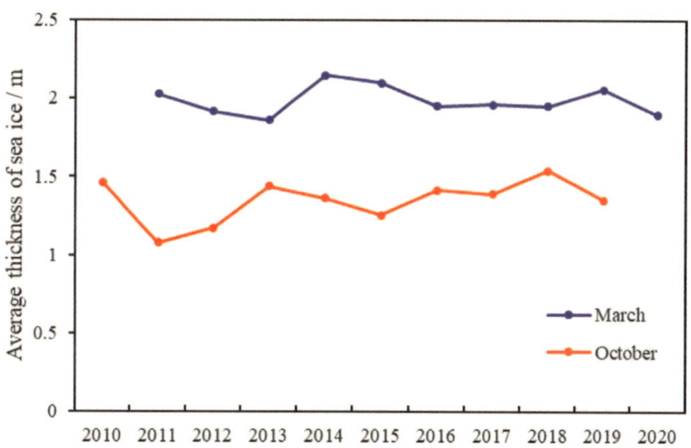

Figure 17.
The average thickness of sea ice in March and November from 2010 to 2020.

4.2 Analysis of volume changes of first-year ice and multiyear ice

To further analyze the process and trend of sea ice change, the volume of first-year ice and multiyear ice was calculated separately over the years. **Figure 18** shows the variation of the sea ice volume in March from 2011 to 2020, and **Figure 19** shows the variation of the sea ice volume in October from 2010 to 2019.

There are small fluctuations in the total sea ice volume in March from 2011 to 2020, And from 2013 to 2014, the total sea ice volume increases by 10.41%. In addition, in 2020, the total sea ice volume decreases by 10.08%, and in the rest of the years, the volumes of the sea ice show some fluctuations with the variation rate being within 5%. Overall, the total volume of sea ice in the Arctic region was relatively stable in recent years. However, by analyzing the variations of the multi-year ice and the first-year ice volume, they appeared opposite trends in recent years. Multi-year ice volume increased by 61.93% in 2014 and 22.70% in 2017, while showing a continuous decreasing trend in 2018 and beyond. First-year ice volume decreased by 12.40% in 2014 and 12.51% in 2020, and increased by 15.53% in

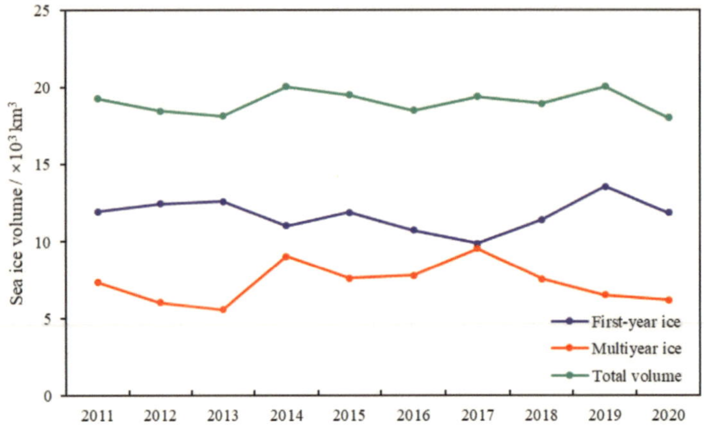

Figure 18.
Folding line graph of sea ice volume change from March 2011 to 2020.

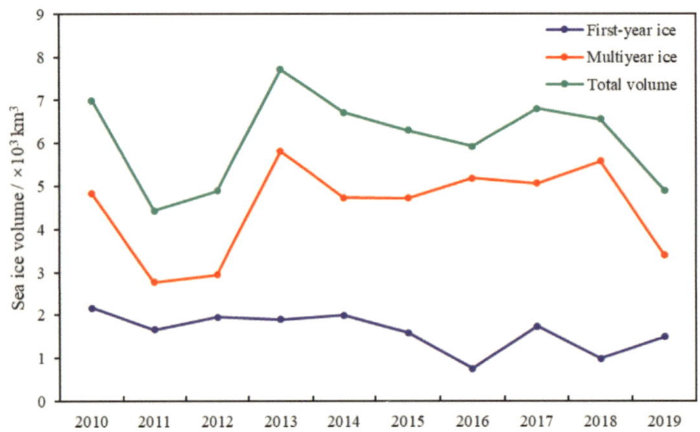

Figure 19.
Folding line graph of sea ice volume change from October 2010 to 2019.

2018 and 18.85% in 2019. The volume changes of the first-year ice and the multi-year ice showed a negative correlation. The growth of one side was accompanied by the decrease of the other side.

From 2010 to 2019, the sea ice volume in October was highly variable. It decreased significantly in 2011 and showed a large increase in 2013. Furthermore, after 2017, the sea ice volume showed a continuous downward trend. Because usually, the life period of the first-year ice would not exist more than one summer, after the summer, the first-year ice maintains a very low total volume in October, and the volume variation of the multi-year ice would be consistent with the volume variation of the total sea ice. The first-year ice volume remained stable in 2014 and before. Except in 2017, the volume of the first-year ice increased to close to the pre-2014, it changed fluctuated less during the rest of the years. As shown in **Figure 18,** the first-year ice volume shows an increasing trend in March 2017. Overall, in recent years, the first-year ice in the Arctic region grew rapidly in winter and melted in summer, while the amount of the first-year ice which became the multi-year ice through the summer decreased.

In summary, from 2010 to 2019, the total volume of the sea ice in March over the years remained stable, which is consistent with the findings of Julienne Stroeve et al. [43]. The total volume of sea ice in October over the years fluctuates less after 2013, but the overall trend is gradually decreasing. However, the total multi-year ice volume showed a decreasing trend [44]. In winter, the increase of the total first-year ice hided the decrease of the total multi-year ice, and the change of the total sea ice volume was not obvious. While in the summer, much first-year ice melted, therefore the increase of the total first-year ice would be hard to compensate for the loss of the total multi-year ice.

5. Conclusion

In this study, the sea ice freeboard and thickness are calculated by the TFMRA algorithm based on the CryoSat-2 satellite L1b altimetry data in the ocean region north of 60°N latitude, including the echo signal preprocessing, waveform classification, and re-tracking correction. Furthermore, the thickness results were made into a grid of 5 km grid resolution and the average sea ice thickness and volume were calculated. From 2010 to 2020, the variation of the average sea ice thickness in March is relatively stable over the years. And there are small fluctuations in the average thickness change of the sea ice in October over the years, while the overall trend is stable. On the other hand, the total sea ice volume showed some small changes in recent years, while tends to decline slowly in general. The first-year ice shows a trend of increasing winter growth and summer ablation with a similar amount of growth and ablation. Decrease in the number of volumes of first-year ice transformed into multi-year ice. The total amount of the multi-year ice changed sharply in the previous years, while showing a slow decreasing trend in recent years.

Our results show that the errors are present in current estimations of Arctic sea ice thickness. Firstly, the threshold q used to calculate echoes is an empirical value, which might introduce some errors in a small range. Secondly, due to the limited amount of Cryosat-2 altimetry data available, the sea ice thickness inversion results exhibit some striped contamination, which may lead to errors. Furthermore, uncertainties arising from changes in snow thickness, sea ice density, snow density, seawater density values, and Arctic circulation patterns can impact the accuracy of sea ice thickness inversions to varying degrees.

Despite these uncertainties, the CryoSat-2 altimetry satellite method offers high accuracy and volume, making it a reliable tool for monitoring Arctic sea ice. In future research, we aim to further investigate the variability of Arctic sea ice thickness for a more comprehensive understanding of this important area.

Conflict of interest

The authors declare no conflict of interest.

Author details

Xubing Zhang[1], Ruizhi Li[1], Wenxi Li[1]*, Liang Zhang[2] and Kai Wang[3]

1 School of Geography and Information Engineering, China University of Geosciences, Wuhan, China

2 State Key Laboratory of Geological Processes and Mineral Resources, School of Earth Sciences, Planetary Science Institute, China University of Geosciences, Wuhan, China

3 School of Geophysics and Geomatics, China University of Geosciences, Wuhan, China

*Address all correspondence to: hmyu_fzdmxtfz@126.com

References

[1] Djepa V. Sensitivity analyses of sea ice thickness retrieval from radar altimeter. Journal of Survival Mapping Engineering. 2014;**2**:44-45

[2] Bouzinac C. CryoSat Product Handbook: ESRIN/ESA and Mullard Space Science Laboratory. London, UK: University College London; 2013

[3] Laxon S, Peacock N, Smith D. High interannual variability of sea ice thickness in the Arctic region. Nature. 2003;**425**(6961):947-950

[4] Peacock NR, Laxon SW. Sea surface height determination in the Arctic Ocean from ERS altimetry. Journal of Geophysical Research: Oceans. 2004; **109**:C7. DOI: 10.1029/2001JC001026

[5] Kwok R, Cunningham GF, Wensnahan M, Rigor I, Zwally HJ, Yi D. Thinning and volume loss of the Arctic Ocean sea ice cover: 2003–2008. Journal of Geophysical Research: Oceans. 2009; **114**(C7):1-16. DOI: 10.1029/2009jc005312

[6] Laxon SW, Giles KA, Ridout AL, Wingham DJ, Willatt R, Cullen R, et al. CryoSat-2 estimates of Arctic Sea ice thickness and volume. Geophysical Research Letters. 2013;**40**(4):732-737. DOI: 10.1002/grl.50193

[7] Rinne E, Similä M. Utilisation of CryoSat-2 SAR altimeter in operational ice charting. The Cryosphere. 2016; **10**(1):121-131. DOI: 10.5194/tc-10-121-2016

[8] Lee S, Im J, Kim J, Kim M, Shin M, Kim H, et al. Arctic Sea ice thickness estimation from CryoSat-2 satellite data using machine learning-based lead detection. Remote Sensing. 2016;**8**(9): 698. DOI: 10.3390/rs8090698

[9] Giles KA, Laxon SW, Wingham DJ, Wallis DW, Krabill WB, Leuschen CJ, et al. Combined airborne laser and radar altimeter measurements over the Fram Strait in May 2002. Remote Sensing of Environment. 2007;**111**(2–3):182-194. DOI: 10.1016/j.rse.2007.02.037

[10] Jain M, Andersen OB, Dall J, Stenseng L. Sea surface height determination in the Arctic using Cryosat-2 SAR data from primary peak empirical retrackers. Advances in Space Research. 2015;**55**(1):40-50. DOI: 10.1016/j.asr.2014.09.006

[11] Helm V, Humbert A, Miller H. Elevation and elevation change of Greenland and Antarctica derived from CryoSat-2. The Cryosphere. 2014;**8**(4): 1539-1559. DOI: 10.5194/tc-8-1539-2014

[12] Kurtz NT, Galin N, Studinger M. An improved CryoSat-2 sea ice freeboard retrieval algorithm through the use of waveform fitting. The Cryosphere. 2014; **8**:1217-1237. DOI: 10.5194/tc-8-1217-2014

[13] Rostosky P, Spreen G, Farrell SL, Frost T, Heygster G, Melsheimer C. Snow depth retrieval on Arctic Sea ice from passive microwave radiometers—Improvements and extensions to multiyear ice using lower frequencies. Journal of Geophysical Research: Oceans. 2018;**123**(10):7120-7138. DOI: 10.1029/2018JC014028

[14] Warren SG, Rigor IG, Untersteiner N, Vladimir R, Nikolay B, Yevgeniy A, et al. Snow depth on Arctic Sea ice. Journal of Climate. 1999;**12**(6): 1814-1829. DOI: 10.1175/1520-0442(1999)012<1814:SDOASI>2.0.CO;2

[15] Kurtz NT, Farrell SL. Large-scale surveys of snow depth on Arctic Sea ice

from operation IceBridge. Geophysical Research Letters. 2011;**38**(20):1-5. DOI: 10.1029/2011GL049216

[16] Cavalieri D, Markus T, Comiso J. Amsr-e/Aqua Daily l3 12.5 Km Brightness Temperature, Sea Ice Concentration, and Snow Depth Polar Grids v002. Boulder, Colorado USA: National Snow and Ice Data Center; 2004. pp. 9017-9024

[17] Stroeve J, Meier W. Sea Ice Trends and Climatologies from SMMR and SSM/I-SSMIS, Version 1. Boulder, Colorado, USA: NASA National Snow and Ice Data Center Distributed Active Archive Center; 2003

[18] Fetterer F, Stewart JS, Meier WN. MASAM2: Daily 4 Km Arctic Sea Ice Concentration, Version 1. Boulder, CO: National Snow and Ice Data Center; 2015

[19] Aaboe S, Breivik L, Sørensen A. Global Sea Ice Edge and Type Product user's Manual. France: EUMETSAT OSISAF; 2016

[20] Brown RD, Brasnett B. Canadian Meteorological Centre (CMC) daily snow depth analysis data. Environment Canada. 2010;**169**:1-18

[21] Krabill WB. IceBridge ATML1B Qfit Elevation and Return Strength. Boulder, CO: NASA DAAC at the National Snow and Ice Data Center; 2010

[22] Kurtz N, Studinger MS, Harbeck J, Onana V, Farrell SL. IceBridge Sea Ice Freeboard, Snow Depth, and Thickness, Version 1. Boulder, CO: NASA National Snow and Ice Data Center Distributed Active Archive Center; 2012. DOI: 10.5067/7XJ9HRV50O57

[23] Gaina C, Werner SC, Saltus R, Maus S. Circum-Arctic mapping project: New magnetic and gravity anomaly maps of the Arctic. Geological Society. 2011;**35**(1):39-48. DOI: 10.1144/M35.3

[24] Skourup H, Farrell SL, Hendricks S, Ricker R, Armitage WK, Ridout A, et al. An assessment of state-of-the-art mean sea surface and geoid models of the Arctic Ocean: Implications for sea ice freeboard retrieval. Journal of Geophysical Research: Oceans. 2017; **122**(11):8593-8613. DOI: 10.1002/2017JC013176

[25] Ridout A. New Mean Sea Surface for the CryoSat-2 L2 SAR Chain. London, UK: UCL; 2014

[26] Feng S, Li F, Li S. Introduction to Marine Science. Beijing, China: Higher Education Press; 1999

[27] Guo J. Satellite Radar Altimetry Waveform Redefinition and Application. Beijing, China: Survey and Mapping Press; 2013

[28] Zhai G, Huang M, Ouyang Y, Lu X. Satellite altimetry principles and their applications. Hydrographic Surveying and Charting. 2002;**22**(1):57-62. DOI: 10.3969/j.issn.1671-3044.2002.01.016

[29] Laxon S. Sea ice altimeter processing scheme at the EODC. International Journal of Remote Sensing. 1994;**15**(4): 915-924. DOI: 10.1080/0143116940 8954124

[30] Thorpe SA. Fluid dynamics, introduction and laboratory experiments. In: Encyclopedia of Ocean Sciences. Cambridge, US: Academic Press; 2001. p. 1070. DOI: 10.1006/rwos.2001.0500

[31] Tilling RL, Ridout A, Shepherd A. Estimating Arctic Sea ice thickness and volume using CryoSat-2 radar altimeter

data. Advances in Space Research. 2018; **62**(6):1203-1225. DOI: 10.1016/j.asr.2017.10.051

[32] Spreen G, Kern S, Stammer D, Hansen E. Fram Strait sea ice volume export estimated between 2003 and 2008 from satellite data. Geophysical Research Letters. 2009;**36**(19):1-6. DOI: 10.1029/2009GL039591

[33] Alexandrov V, Sandven S, Wahlin J, Johannessen OM. The relation between sea ice thickness and freeboard in the Arctic. The Cryosphere. 2010;**4**:373-380. DOI: 10.5194/tc-4-373-2010

[34] Golden KM, Ackley SF, Lytle VI. The percolation phase transition in sea ice. Science. 1998;**282**(5397):2238-2241. DOI: science.282.5397.2238

[35] Lytle VI, Ackley SF. Heat flux through sea ice in the western Weddell Sea: Convective and conductive transfer processes. Journal of Geophysical Research: Oceans. 1996;**101**(C4): 8853-8868. DOI: 10.1029/95JC03675

[36] Shen S. A Study of CrvoSat-2 Based Sea Ice Thickness Inversion Method. Research on Sea Ice Thickness Inversion Method Based on CrvoSat-2. Nanjing: Nanjing University; 2018

[37] Lewis MJ, Tison J, Weissling B, Delille B, Ackley SF, Brabant F, et al. Sea ice and snow cover characteristics during the winter–spring transition in the Bellingshausen Sea: An overview of SIMBA 2007. Deep Sea Research Part II: Topical Studies in Oceanography. 2011; **58**(9–10):1019-1038. DOI: 10.1016/j.dsr2.2010.10.027

[38] Farrell SL, Kurtz N, Connor LN, Elder BC, Leuschen C, Leuschen C, et al. A first assessment of IceBridge snow and ice thickness data over Arctic Sea ice. IEEE Transactions on Geoscience and Remote Sensing. 2011;**50**(6):2098-2111. DOI: 10.1109/TGRS.2011.2170843

[39] Jiao Y, You Q, Li Y, Zhou Y. Seasonal and interdecadal variability of Arctic Sea ice and temperature response in East Asia. In: 35th Annual Meeting of the Chinese Meteorological Society S7 East Asia Climate, Variability Mechanisms of Extreme Climate Events and Climate Prediction. Vol. 11. Beijing, China: Chinese Meteorological Society; 2018. p. 423

[40] Kwok R, Rothrock DA. Decline in Arctic Sea ice thickness from submarine and ICESat records: 1958–2008. Geophysical Research Letters. 2009; **2009**:36. DOI: 10.1029/2009GL039035

[41] Sumata H, de Steur L, Divine DV, et al. Regime shift in Arctic Ocean sea ice thickness. Nature. 2023;**615**(7952): 443-449. DOI: 10.1038/s41586-022-05686-x

[42] Xiao F, Zhang S, Li J, et al. Arctic Sea ice thickness variations from CryoSat-2 satellite altimetry data. Science China Earth Sciences. 2021;**64**(7):1080-1089. DOI: 10.1007/s11430-020-9777-9

[43] Stroeve J, Holland MM, Meier W, et al. Arctic Sea ice decline: Faster than forecast. Geophysical research letters. 2007;**34**(9). DOI: 10.1029/2007GL029703

[44] Pörtner HO, Roberts DC, Masson-Delmotte V, et al. The Ocean and Cryosphere in A Changing Climate. IPCC Special Report on The Ocean and Cryosphere in a Changing Climate2019. p. 1155. DOI: 10.1017/9781009157964

Chapter 3

The Nexus of Climate Change, Urban Infrastructure and Sustainable Development in Developing Countries

Olayinka C. Oloke and Nelson A. Akindele

Abstract

This study weighed in on the topical issue of climate change disasters and sustainable urbanization in developing countries. Climate change is a global challenge and burning discourse at many national and international fora. The study ruminates on the lot and plight of cities and peri-urban communities in developing countries amidst the rising concern over climate change pattern and hazards. Scoping review of relevant literatures was used for the research method. Specifically, studies relating to the basic concept, that is climate change, urban infrastructure and sustainable development were reviewed qualitatively in order to present the viewpoint of the current study. The study scrutinized the threats of climate change across the globe as well as the implications for dwellers in developing countries like Nigeria. The process of urbanization and the state of infrastructure were also highlighted. Vulnerability of urban settlements in terms of infrastructure shortage was brought to the fore in order to enhance their preparedness and capacity to withstand the pangs of extreme weather events. The study concluded by advocating positive urbanization in developing countries which is the essence of sustainable development of human settlements.

Keywords: infrastructure, climate, sustainable, urbanization, developing

1. Introduction

The rising spate of phenomenon and disaster associated with the global warming and resultant extreme weather events in the 21st century have beamed attention on climate change in most national and international discourse. Across the world, the rising incidence of various environmental, ecological and atmospheric disruptions such as extreme heat, heat wave, wildfire, heavy winds, hurricane, heavy rain, storm surge, flooding, rising sea level, snow storm, landslide, drought, etc. have been largely attributed to the unusual fluctuations of weather elements and the climatic system. The havocs wrecked by these events endanger humanity, environment and economic system of nations and as a result, reports oftentimes present a universal dimension of the phenomenon. According to the World Bank [1], climate change portends severe dangers to all living things, the entire ecosystem and infrastructure. Furthermore, countries around the world are persistently affected by the unusual patterns of dangerous climate events. Apart from the qualitative description of the magnitude, the phenomenon has also been buttressed with scientific proof and quantitative assessment of the impact [2], averred that prior to industrial revolution, the amount of atmospheric carbon-dioxide (CO_2) was within 260–280 part per million by volume (ppmv), but have risen by about 36 per cent to about 380 ppmv in the last (20th) century. In the same manner [3], found that since the arrival of industrialization, the concentrations of methane (CH_4) and CO_2 have risen by 70 per cent beginning from 1970 to 2004. The [4] concluded that the rise in global temperature often referred to as global warming is real. The report further affirmed that the mean earth heat has increased by 0.76°C and sea level has risen by 17 cm in nineteenth century. Likewise, some world's leading climate scientist in 2008, strongly advocates urgent reduction of CO_2 emission [2] to forestall calamity. According to the experts, "*paleoclimate* signal and current climate fluctuations suggest that there is need for reducing atmospheric carbon-dioxide from its current 385 ppmv to at most 350 ppmv, else, the consequences would be irremediable tragedy [5].

The gap between the elite and the poor is getting wider as a result of the havocs of the climate change. As the hazards strains the financial resources of the rich, it further impoverish the poor, thereby aggravating their vulnerability. As often revealed in reports, the cost of climate fluctuations calamity across the globe is usually enormous. For instance, flooding in Sante Fe, Argentina displaced one-third of the city population and affected 27,928 households [6]. Damages were approximated at US$ 1 billion: US$ 752,000,000 in manufacturing, animal husbandry, agriculture and trade sectors; US$ 180,000,000 in infrastructure damages while societal loss was put at US$ 91,000,000 [6]. In Nigeria, the destructive pattern of inclement weather elements have aggravated socio-economic and infrastructure damage and loss. [7] recorded that an estimate of US$ 45 million was lost annually in twelve states due to erratic heavy winds, heavy rain and storm surge in 16 years. By implication, the sum of US$ 2.2 billion was lost across the thirty-six states and the federal capital territory within the period. The report further showed that the overall financial loss as a result of the rising precipitation and heavy wind rose from US$ 23.6 million to US$ 82.2 million between 1992 and 2007, with the highest of USD91.5 million in 2006 and lowest of US$ 17.97 million in 1994 respectively. Across the globe, coastal cities are witnessing unprecedented rise in sea level, massive flooding, horrendous storm/hurricane, severe erosion, saltwater incursion and disruption in sedimentation formation [8]. It was further revealed that tropical gales upset 1.4 billion people, which constitutes 24 percent of the riverine settlers. Furthermore, in 2010, floods tragedy in Pakistan and Chile affected about 18,102,237 and 134,000,000 people, caused 1985 and 1691 deaths and damages estimated at US$ 9.5 billion and US$ 18 billion respectively [9]. Hurricane Katrina that ravaged New Orleans in 2005 also caused the death of 1833 persons, displacement of 500,000 people and a loss of US$ 125 billion. In Senegal, the World Bank estimates the value of assets exposed to flooding at about €40 billion (Forty billion Euro), twice the Gross Domestic Product of the country [10]. According to [11], over the past ten years, climate change has caused about 3852 catastrophes, the death of more than 780,000 and displaced over 2 billion people with property worth more than US$ 960 billion lost to the tragedy.

Arising from the above, the reality of climate change and its huge financial implications of the consequences become very conspicuous. Reports across the world according to [12] showed disturbing pattern of climate change consequences.

Nonetheless, public discourse generally has attributed disparate level of urgency to issues of climate change thereby prioritizing it above other critical economic and environmental debacles that threaten human existence. Davenport et al. [13] stated that up to 70 member states of the United Nations significantly prioritize climate change risk above security risk. Nations have essentially ascribed the various socio-economic, ecological, environmental, political and security problems to climate change. [14] observed that most countries face economic, political, and social hardship resulting from uncertainties occasioned by climate change hazards. According to [15], countries like Turkey, Brazil, Egypt and Iraq are confronted with security risk triggered by the conflict among armed forces due to the consequences of climate change. A notable example of security challenge of continental significance is the herders' militancy in Nigeria whose violent clashes with farmers have always claim lifes and properties.

Whilst not playing down the danger that climate change poses, it is important to know that the spate of havocs wrecked by climate change events in recent times is as a result of aggressive industrialization and urbanization in the last two centuries. Besides, there are numerous challenges in rural and urban areas around the world that have undermined human dignity and normal/decent existence over the years. These challenges have been aggravated in recent times by climate change. The diverse socio-economic, political, environmental and security challenges in different countries are first and foremost, symptoms of policy failure, bad governance, mismanagement and lack of political will especially in developing countries. These consequently get complicated by the erratic climate manifestations. In this regards, the universal perspective of climate change hazards tends to obfuscate the experience at regional and local levels. Therefore, prioritizing climate change problems over other basic societal challenges engender the risk of mis-characterization, misplaced focus, waste of scarce resources, death and damages to properties and environment. Certainly, climate change according to [8] is intricately connected to urbanization, but not the most severe environmental issues facing humanity. Hardoy and Pandiella [16] concluded that climate change has put the already vulnerable urban settlements under enormous stress thereby making the situation require quick and comprehensive approach to resolve the problems. The socio-economic and environmental problems facing developing nations include acute poverty, widespread illiteracy, insecurity, gender inequality, environmental pollution, infrastructure deficit, unemployment and poor governance to mention a few. In the light of the foregoing, it becomes imperative to put in context the diverse challenges that confront developing countries and develop a comprehensive framework to resolve them. Finding solutions to these issues engender the resilience of human settlements against the impact of climate change in Nigeria and other developing countries.

2. Theoretical framework: theory of change

This study is anchored on the postulations of the theory of change as explained in [17, 18]. The theory of change according to [19] has a wide range of possible uses in developing, managing and evaluating interventions that are meant to produce a desired long term change. Reinholz and Andrews [20] described a set of assumptions that should be taken to achieve a long term goal as well as the linkages between the activities and results of an intervention program. Davies [21] stated that theory of change is a process that is meant to evolve an articulated sequence of events that will eventually usher in the expected outcome. The study further described the theory as a strategic map that shows the multiple interventions required to produce immediate results and are pre-conditioned to a long time change. It depicts the pathway that is consistent to producing the impact expected to support the desired long term change. In the application of the theory of change, [22] suggest that to accomplish a successful change, there is need to empower the driving forces and weaken the restraining forces. This according to [22] could be achieved through the following:

i. intensifying the driving forces responsible for diverting the behavior away from the present situation;

ii. reducing the influence of forces that adversely affect the movement away from the existing situation; or

iii. combining the two methods

The relevance of the theory of change to the current study is contained in the inherent popular interest to tackle the environmental and ecological hazards unleashed by global warming. The incidence of climate change has been debated in the past decades, however, the reality is being established by the ominous signs of its impacts globally. Notwithstanding, the hype about the phenomenon in developing countries distract attention from issues that are grave, grievous and constitute daily threat to human existence in this part of the globe. As such the study believe in prompt intervention as enunciated in the theory of change and align with [22] to empower the driving forces and weaken the restraining forces especially for the developing nations. This could be achieved by developing appropriate policy and developmental interventions in the areas of urbanization process and infrastructure delivery based on the principles of sustainable development. This would ultimately reinforce the capacity to withstand and combat the periodic barge of climate change in less advanced countries.

3. Methodology

Scoping review was adopted as the research method for undertaking this study. This approach was engaged due to the broad and intricate nature of the specific concepts involved. A scoping review of a body of literature can be of particular use when the topic has not yet been extensively reviewed or is of a complex or heterogeneous nature [23]. Hence, studies and reports on the subject of climate change, urban infrastructure and sustainable development were scoped in the review process. A scoping review identifies the key concepts in a research area together with the main sources and types of evidence. The strategy was engaged to identify relevant body of literature, put the current research in context, summarize and present research findings and make necessary recommendations.

4. Urbanization challenges in Africa

In the preceding sections, it was revealed that climate change is one of the negative outcomes of intensified urbanization. Urbanization depicts increase in the number of urban dwellers. Usually, urbanization is as a result of rise in natural birth rate, migration to urban areas and regrouping of rural into urban areas. However, industrialization has transformed the process of urbanization around the world. Consequently, this has redefined the overall perception of urbanization. According to [24], urbanization is the transformation of rural to urban centres via the process of economic enhancement and mechanization. This implies that urbanization is a transition from primarily rural settlement to urban centre. Through proper planning, good governance and efficient management of human, capital and natural wealth, urbanization heralds opportunities for socio-economic growth and development. In addition, it brings employment opportunities and provide platform for sustainable urban development. Urbanization in many developing nations is driven by natural birth and rural–urban migration. This though has been rapid but not supported with investment in infrastructure, capacity development and entrepreneurship. As a result, the process has not resulted in liability for the developing nations. As pointed out in [8], urbanization in Africa has not resulted in economic transformation and prosperity as expected. Illiteracy, poor physical infrastructure and bad governance have hindered the efficient use of productive resources to promote economic growth and development. Current economic momentum in Africa according to [8] is tied to the prosperity in other regions of the world. Despite the surge in urban dwellers in many African countries in the past decades, diverse economic and environmental challenges are still predominant ranging from infrastructure deficit, pollution, congestion, inadequate shelter, and poverty [24]. The haphazard urbanization has caused acute socio-economic, cultural and environmental issues for the developing nations [25–27].

Nigeria is the most populous nation and one of the rapidly urbanizing nations in sub-Sahara Africa. In Nigeria, as obtained in many African countries, urbanization predates colonization and industrialization in the 19th century, when economic activities such as trading and marketing and administrative expediency played critical roles in the growth and development of urban settlements [28]. In the 20th century, the growth in urban population and spread of urban centres were attributed to distinct economic cum policy developments which include the development of wheeled transportation, classification of settlements by hierarchy, transition to monetized economy, periodic geopolitical restructuring and lately, industrialization process between 1960 and 1975 [29]. In addition, the oil boom in the 70s accelerated infrastructure provision in major urban centres and caused remarkable rural–urban pull in Nigeria. Consequently, urban growth in Nigeria have been swift, spontaneous and uninhibited [30] leading to all sorts of haphazard expansion and infrastructure collapse. Studies [31–33] have shown that poor planning and intense land use in the inner-city has exacerbated urban problems. Upholding this, [34, 35] averred that the rate and style of development in Nigeria has caused greater harm to the country. The problems as mentioned in the studies include slum proliferation, pollution, violence and crime, environmental degradation, sicknesses, poverty and complex traffic challenge. This situation is similar in most urban centres in Nigeria. For instance, [36] maintained that Ibadan city is beset with diverse problems which are mostly consequences of population growth, macro-scale economic conditions, environmental challenges and weak urban development policy. Consequently from the foregoing, the following points came to the fore:

i. urbanization in developing countries precedes industrialization

ii. direct relationship exist between urbanization and urban challenges in developing countries

iii. urbanization in developing countries do not necessarily translate to industrialization and economic empowerment of the people

5. Urban infrastructure in Africa

The challenges of urban infrastructure in developing countries are enormous, complex and intertwined. These often range from complete lack to inadequate infrastructure and decay of existing infrastructure. Urban centres across Africa suffer a great deal from infrastructure deficiency and this constitutes formidable hindrance to national economic growth. Across the world, [37] stated that an estimated one billion city dwellers occupy slums. Moreover, around fifty percent of city populace in sub-Saharan countries does not have access to basic sanitation while close to twenty percent lack access to safe water, good drainage and proper waste management [37]. Generally, there is acute shortage of basic infrastructure such as housing, pipe borne water, drainage, sanitation, solid and waste water treatment facilities, transport, rail and road infrastructure, power and health infrastructure [8]. [38] also averred that cities in developing countries are bedeviled with dysfunctional economic system, infrastructure deficiencies, governance failure, weak policies and social breakdown. Despite these, physical expansion or rather, urban sprawl continues uninhibited without commensurate growth in economic activities and infrastructure development thereby culminating in a strong positive correlation with the rise in societal disorderliness and diverse challenges. Unlike the nature of urban growth experienced in advanced countries which was equally driven by population increase among other factors, urban growth in developed countries occur hand in hand with economic growth [39]. In those countries, the issue there is more of maintenance of infrastructure than lack of it. Good governance and strong economic development policies have been put in place to propagate positive urbanization and sustainable physical development. On the contrary, the rising poverty index among the urban populace is an indication of persisting economic conundrum facing most developing countries. For example, the official statistics of Nigeria's poverty profile revealed that the relative poverty index rose from 54.4% in 2004 to 69% in 2010 [40]. Road network experience heavy traffic almost every day and do not have pedestrian walkway, flood/waste water canal and are poorly maintained [40]. The sub-urban communities usually lack good road networks and are difficult to access during raining season. Pipe borne water is basically non-existent even in city centres and many households cannot afford proper treatment of their solid and waste water. There is the proliferation of slums in urban areas and shortage of housing units. The current state of infrastructure in urban centres of most developing countries is in a dire strait and is again being pummeled by the occasional rage of climate change.

6. Economic growth and infrastructure development in Nigeria

The state of economic growth in Nigeria has been on the decline in the last two to three decades, throwing millions deeper down the pit of poverty. This though, is not unconnected to the population growth, bad leadership, poor policies and endemic corruption. Studies have shown that rapid population growth slows down the growth of per capital income and perpetuate inequalities of income distribution in developing countries [41]. Before the COVID-19 crisis 2020, it was estimated that about 4 in 10

Nigerians were living in poverty and millions more were vulnerable to falling below the poverty line [41]. However, based on the most recent official survey data from the Nigerian National Bureau of Statistics, 39.1 percent of Nigerians lived below the international poverty line of $1.90 per person per day (2011 PPP) in 2018/19 [42]. Yet a further 31.9 percent of Nigerians had consumption levels between $1.90 and $3.20 per person per day, making them vulnerable to falling into extreme poverty when shocks occur. The continued dependence on oil, high population growth rate and limited job creation has hampered the broad-based growth required to tackle poverty. In the same vein, the Nigeria's energy poverty has been quite appalling [43, 44]. Currently, Nigeria has an approximate population of over 200 million with an average estimated demand of 31.2 GW, and the country has an installed capacity of 14.38 GW with an average supply capacity of 6GW [45]. This implies that the system could only generate 40% of its installed capacity. This acute energy deficit has resulted in economic decline, low standard of living, citizen hardship and widespread poverty. It has scared foreign investors and forced several industries to relocate to countries with much more reliable and stable energy supply. The economic impacts and loss of human capital occasioned by the state of electricity infrastructure in Nigeria, is by all ramifications, far greater than the loss induced by climate change hazards [45]. [46] and [47] concluded that stable and reliable energy supply plays significant role in the economic development, poverty reduction industrial, agriculture, manufacturing, commerce, infrastructure development, employment, and security. In addition, it plays a vital role in ensuring that basic needs and services (food and water, housing, health services, and education) are adequately provided [48]. Furthermore, medical tourism and declining standard of education are another consequence of a broken governance system with severe impact on the citizenry. Medical tourism not only manifest in citizens seeking the best of healthcare in India, Europe and United States, but also in brain drain of health workers, particularly the nurses and medical doctors to countries where their services are adequately remunerated. It was reported in [49] that about N576 billion ($1.2 billion) is lost to medical tourism yearly in Nigeria. Corroborating this, Price Waterhouse Coopers report that Nigerians spend $1 billion annually on medical tourism [49]. [50] listed brain drain, underfunding, dilapidated structures and obsolete equipment, industrial strike, and negative attitude of health professionals as problems facing the Nigerian health sector. These and several other socio-economic problems have contributed more to the economic backwardness and infrastructure disease that plaque the Nigeria nation. These were not as a result of climate change, but only compounded by climate change disasters.

7. Sustainable urban development

Industrialization, urbanization process and man's anthropogenic activities have increased atmospheric temperature and triggered the phenomenon known as global warming. This in turn has culminated into unpredictable pattern of climate events/disasters in recent times. Climate change and urbanization are both global phenomenon though the impact are largely localized. By implication, climate change can release its disrupting capability on any part of the earth. The advanced nations have contributed more than the developing countries, to the emission of greenhouse gases that cause global warming. The unpredictable patterns of weather events triggered by global warming often unleash tragedies at places without regard to the level of development, type or pattern of settlement. However, the factors that distinguish locations lie in their resilience and recovery capability. Urban resilience has to do with the level of economic and infrastructure development while recovery is contingent on good policy, proper planning and good governance. According to [8], urban population will record 95% growth over the next two decades in developing countries. The report admitted that the dramatic rise in urban population of emerging nations exposes the masses and infrastructure to dangerous environmental consequences. Climate change disasters expose the susceptibility of built-up neighborhoods in developing countries and compound the existing urban challenges. **Table 1** presents the global occurrence of natural hazards between 2002 and 2010.

The Table showed relevant statistics that reinforce the position of this study. The frequency, mortality rate and affected locations all points to the fact that climate change disasters are usually localized. As shown, about 228 earthquake disasters happened across the globe between 2002 and 2010. Out of this, only seven occurred in Africa [51]. Countries affected include DR Congo in 2002 & 2008; Algeria in 2003; Morocco in 2004; Tanzania in 2005; Mozambique in 2006 and Malawi in 2009. Similarly, the record of volcanic eruptions globally in the 21st century showed that only Democratic Republic of Congo and Eriteria were affected in Africa in the year 2002 and 2011 respectively [52]. Furthermore, the incidence of drought, flooding and heat wave in inner cities and northern parts are being managed with modern technology and infrastructure. Unlike other economic and environmental plagues that has become recurring decimal in Nigeria and many Africa countries.

Classification of hazard	Nature of hazard	Frequency of occurrence	Mortality risk	Vulnerable urban areas
Geophysical	Earthquake	228	High	Cities on or near fault lines
	Volcano	53	Low	Cities near volcanoes
Geophysical and hydrological	Tsunami	19	High	Coastal cities
	Mass movement (landslide, rockfall, avalanche, subsidence)	167	Low	—
Hydrological	Flood	1501	Medium	Coastal cities
	Storm surge	25	Low	Coastal cities
Meteorological	Storm and cyclone	899	High	Coastal cities tropical cities
Climatological	Drought	133	Low	Cities in or near desert & dry areas
	Extreme temperature (heat and cold)	173	Medium – high	Inland cities
	Wildfire	101	Low	—

Source: EMDAT, OFDA/CRED International Disaster Database (2011) cited in [8].

Table 1.
Global occurrence of natural hazards between 2002 and 2010.

Amidst the predicament of urbanization and infrastructure deficit, the concept of sustainability has been introduced in the present and future development of man's physical environment and almost every aspect of human activities. [53] described sustainable development as one that satisfies current needs without jeopardizing that of future generations. In order not to expend limited resources on finding solution to existing urban misnomer, the concept of sustainable development plays an important role. Sustainable development is crucial to re-direct and redefine urbanization in developing countries. Taking a cue from the United Kingdom, the UK political agenda has centered on sustainability agenda in the last three decades [54]. Sustainable development has the remedy to the existing urbanization pattern in developing countries. Sustainable development combines infrastructure development with economic growth and employment opportunities thereby converting population growth to economic advantages. For instance, the UK Government's sustainable development agenda for residential development is based on four concurrent objectives ([55] cited in [54]):

- Social progress which meets the needs of everyone;
- Effective protection of the environment,
- Prudent use of natural resources, and
- Maintenance of high and stable levels of economic growth and employment

The UK department of Community and Local Government (CLG) recognizes that sustainable community should comprise "*…places where people want to live and work, now and in the future, meet the diverse needs of existing and future residents, are sensitive to their environment and contribute to a high quality of life. Such communities are safe and inclusive, well planned, built and run and offer equal opportunity and good services for all*" [56].

8. Conclusion and recommendation

Without doubt, climate change is happening as claimed in different studies earlier cited. Nevertheless, the phenomenon is certainly not the most critical problem facing developing countries at this time. Climate change threatens the fabrics of every settlement alike, but of higher significance is the negative urbanization syndrome that characterizes urban growth in developing countries. There is therefore the urgent need to reverse the trend of urbanization in most African countries to embrace positive urbanization. Positive urbanization connotes economic growth that combines infrastructure development with employment opportunities for the rising population. Positive urbanization is the primary intent and result of sustainable development of urban centres. It is an inclusive development strategy that engenders disaster preparedness of neighbourhoods and goes further to absorb and absolve the problems of previous urbanization anomalies. Consequently, positive urbanization, whether driven by natural birth, rural–urban migration, industrialization or combination of all is *sine qua non* for meaningful sustainable urban development in developing countries. Through this, investment in infrastructure is prioritized, right policies are enacted to empower the masses, conducive atmosphere is created for entrepreneurship to thrive thereby creating employment opportunities. Sustainable urban development engenders urban resilience and improves recovery in case of disaster occurrence. Tackling infrastructure gap in urban centres and promoting sustainable development would go a long way to achieving the United Nations' sustainable development goals #10, #11 and #13 in developing countries.

Acknowledgements

The authors wish to acknowledge the painstaking scrutiny and constructive contributions of reviewers and editors. The support received from Covenant University Centre for Research, Innovation and Development is equally appreciated.

Author details

Olayinka C. Oloke[1*] and Nelson A. Akindele[2]

1 Department of Estate Management, College of Science and Technology, Covenant University, Ota, Ogun State, Nigeria

2 Durham School, College of Engineering, University of Nebraska–Lincoln, United States

*Address all correspondence to: yinka.oloke@covenantuniversity.edu.ng

References

[1] World Bank. Cities and Climate Change: An Urgent Agenda. A Report of the International Bank for Reconstruction and Development Bank and World Bank; World Bank. Forward by Inger Andersen. 2010

[2] Litman T. Climate Change Emission Valuation for Transportation Economic Analysis. Victoria, British Columbia, Canada: Victoria Transport Policy Institute; 2012. Available from: www.vtpi.org

[3] United Nations Human Settlements Programme; UN-Habitat. Cities and Climate Change Global Report on Human Settlements. London: Earthscan; 2011

[4] IPCC. Mitigation of Climate Change, Contribution of Working Group III to the Fourth Assessment Report of the Intergovernmental Panel on Climate Change. Cambridge, United Kingdom: Cambridge University Press and New York, NY, USA; 2007

[5] Hansen J, Sato M, Kharecha P, Beerling D, Berner R, Masson-Dalmotte V, et al. Target atmospheric CO_2: Where should humanity aim? The Open Atmospheric Science Journal. 2008;**2**:217-231. DOI: 10.2174/1874282300802010217

[6] Natenzon C. Inundaciones catastrófi cas, vulnerabilidad social y adaptación enun caso argentino actual. Cambio climático, elevación del nivel medio del mar y sus implicancias. In: Paper Submitted to EMF Workshop IX: Climate Change Impact and Integrated Assessment, July 28–August 7, Snowmass, Colorado: EMF Workshop; 2006

[7] Odjugo POA. Quantifying the Cost of Climate Change Impact in Nigeria: Emphasis on Wind and Rainstorms, Journal of Human Ecology. 2009;**28**(2):93-101. DOI: 10.1080/09709274.2009.11906223

[8] Baker JL, editor. Climate Change, Disaster Risk and the Urban Poor: Cities Building Resilience for a Changing World. A Report of the International Bank for Reconstruction and Development and World Bank; Switzerland: World Bank, Urban Development Series; 2012. p. 68358

[9] IFRC. World Disaster Report: Focus on Urban Risk. A Report by the International Federation of red Cross and red Crescent Societies. Geneva: IFRC) Available from: www.ifrc.org; 2010

[10] Saghir J, Santoro J. Urbanization in sub-Saharan Africa: Meeting challenges by bridging stakeholders. In: A Report Produced by Center for Strategic and International Studies; Project on Prosperity and Development. Washington D.C: Center for Strategic and International Studies; 2018

[11] Duran S, Ergun Ö, Keskinocak P, Swann JL. Humanitarian logistics: Advanced purchasing and pre-positioning of relief items. In: Handbook of Global Logistics. New York: Springer; 2013. pp. 447-462

[12] Ahmed N, Khan TI, Augustine A. Climate change and environmental degradation: A serious threat to global security. European Journal of Social Sciences Studies. 2018;**3**(1):161-172. DOI: 10.5281/zenodo.1307227

[13] Davenport C, Gillis J, Chan S, Eddy M. Inside the Paris Climate Deal. Vol. 12. New York Times; UK: Oxford University Press; 2015. p. 15

[14] Schäfer PJ. The concept of security. In: Human and Water Security in Israel and Jordan. Berlin Heidelberg: Springer; 2013. pp. 5-18

[15] Aronoff K, Gindin S, Aschoff N, Karp M, Geismer L, Heideman P. The War on Climate Change. Jacobin. Benin City, Nigeria. 2016. Available from: https://www.jacobinmag.com/2015/11/cop-21-paris-climate-change-global-warming-keystone-pipeline/

[16] Hardoy J, Pandiella G. Urban poverty and vulnerability to climate change in Latin America. Environment and Urbanization. 2009;**21**(1):203-224. DOI: 10.1177/0956247809103019

[17] Funnell S, Roger P. Purposeful Program Theory: Effective Use of Theory of Change and Logic Models. San Franscisco, CA: Jossey-Bass. 2011

[18] Mayne J, Johnson N. Using theory of change in the agriculture for nutrition and health CGIAR research program. Evaluation. 2015;**21**(4):407-428

[19] Mayne J. Useful theory of change models. Canadian Journal of Program Evaluation. 2015;**30**(2):119-142. DOI: 10.3138/cjpe.230

[20] Reinholz DL, Andrews TC. Change theory and theory of change: What's the difference anyway? IJ STEM Ed. 2020;**7**(2):1-12. DOI: 10.1186/s40594-020-0202-3

[21] Davies R. Planning evaluability assessments: A synthesis of the literature with recommendations. Working Paper 40. London: UK Department for International Development; 2013. Available from: https://www.gov.uk/go%0Avernment/uploads/system/uploads/attachment_data/file/248656/wp40-planning-eval-assessments.pdf%0A

[22] Shirey M. Lewin's theory of planned change as a strategic resource. The Journal of Nursing Administration. 2013;**43**(92):69-72

[23] Mays N, Roberts E, Popay J. Synthesizing research evidence. In: Fulop N, Allen P, Clarke A, Black N, editors. Studying the Organisation and Delivery of Health Services: Research Methods. London: Routledge; 2001. pp. 188-219

[24] Momoh J.U. Sustainable Urbanism and its Assessment in Developing Countries: The Nigerian Case [Unpublished Thesis]. Nottingham Trent University; 2016

[25] United Nations. World Population Policies New York. USA: United Nations; 2004

[26] UNCHS. Urbanization: A Turning Point in History. Global Report on Urbanisation; UN-Habitat. 2007. Available from: www.unhabitat.org

[27] UNFPA. State of the World Population. Unleashing the Potential of Urban Growth. New York: United Nation Population Fund; 2007

[28] Asiyanbola RA. Urbanization, gender and transport research issues and insights in Nigeria: Towards a sustainable gender sensitive transport development. The Nigerian Journal of Sociology and Anthropology. 2012;**8**:30-44

[29] Oyesiku OK. Policy framework for urban motorcycle public transport system in Nigerian cities. In: Godard X, Fatonzoun I, editors. Urban Mobility for all. Lisse: A. A. Balkema; 2002. pp. 255-261

[30] Olanrewaju DO. Town planning: A veritable means for poverty reduction. Inaugural Lecture Series 38, delivered at

The Federal University of Technology, Akure on 26th October, 2004

[31] Egunjobi L. Our Gasping Cities. Inaugural Lecture delivered at the University of Ibadan on Thursday, 21st October 1999

[32] Egunjobi LX. Planning the Nigerian cities for better quality of life. In: Onakomaiya SO, Oyesiku OO, editors. Environment, Physical Planning and Development in Nigeria. Ago-Iwoye, Nigeria: Department of Geography and Regional Planning, Olabisi Onabanjo University; 1999. pp. 89-107

[33] Foundation for Urban Development in Africa. The Legacy of Akin Mabogunje. Washington DC, USA: The Cities Alliance; 2006

[34] Aluko OE. The impact of urbanization on housing development: The Lagos experience, Nigeria. Ethiopian Journal of Environmental Studies and Management. 2010;**3**(3):64-74

[35] Adetunji MA, Oyeleye OI. Evaluation of the causes and effects of flood in Apete, Ido local government area, Oyo state, Nigeria. Journal of Civil and Environmental Research. 2013;**3**(7):19

[36] Adelekan IO. Ibadan City Diagnostic Report. Ibadan, Nigeria: UK Department for International Development; #4 for Urban Africa Risk Knowledge. 2016. pp. 1-21

[37] WHO/UNICEF. Progress on sanitation and drinking water. A publication of the World Health Organization and UNICEF Joint Monitoring Programme for Water Supply and Sanitation Update 2010

[38] Rakodi C. African Towns and Cities: Power Houses of Economic Development or Slums of Despair? Paper to "City Future" Conference. Chicago: University of Illinois; 2004

[39] Cohen B. Urban growth in developing countries: A review of current trends and caution regarding existing forecasting; National Research Council, Washington, DC, USA. World Development. 2003;**32**(1):23-51

[40] Nigeria Bureau of Statistics. Social Statistics in Nigeria. Nigeria, Abuja: National Bureau of Statistics; 2012

[41] Adewole AO. Effect of population on economic development in Nigeria: A quantitative assessment. International Journal of Physical and Social Sciences. 2012;**2**(5):1-14

[42] World Bank Group. Poverty and Equity Brief. Africa Western and Central Nigeria: World Bank Group; 2021. Available from: https://databank.worldbank.org/data/download/poverty/

[43] Edomah N, Ndulue G, Lemaire X. A review of stakeholders and interventions in Nigeria's electricity sector. Heliyon. 2021;**7**(9):e07956-e07956. DOI: 10.1016/j.heliyon.2021.e07956

[44] Ebhota WS, Tabakov PY. Power inadequacy, the thorn in economic growth of Nigeria. International Journal of Applied Engineering Research. 2018;**13**(16):12602-12610

[45] Akuru UB, Onukwube EI, Okoro IO, Obe ES. Towards 100% renewable energy in Nigeria. Renewable and Sustainable Energy Reviews. 2017;**71**:943-953

[46] Rafindad AA. Does the need for economic growth influence energy consumption and CO2 emissions

in Nigeria/ evidence from the innovation. Accounting test. Renewable and Sustainable Energy Reviews. 2016;**62**:1209-1225

[47] Ayodele TR, Ogunjuyigbe ASO. Increasing household solar energy penetration through load partitioning based on quality of life: The case study of Nigeria. Sustainable Cities and Society. 2015;**18**:21-31

[48] Chanchangi YN, Adu F, Ghosh A, et al. Nigeria's energy review: Focusing on solar energy potential and penetration. Environment, Development and Sustainability. 2022. DOI: 10.1007/s10668-022-02308-4

[49] Guardian, Nigeria loses over N576b yearly to medical tourism. 2021 https://guardian.ng/news/nigeria-loses-over-n576b-yearly-to-medical-tourism/. [Accessed: June 26, 2022]

[50] Abubakar M, Basiru S, Oluyemi J, Abdulateef R, Atolagbe E, Adejoke J, et al. Medical tourism in Nigeria: Challenges and remedies to health care system development. International Journal of Development and Management Review. 2018;**13**(1):223-238

[51] Meghraoui M et al. The Seismotectonic map of Africa by Mustapha Meghraoui and the ICGP – 601 working group. Episodes. 2016;**39**(1):9-18. DOI: 10.18814/epiiugs/2016/v39i1/89232

[52] List of Volcanic Eruptions in the 21st Century. Available from: https://en.wikipedia.org/wiki/List_of_volcanic_eruptions_in_the_21st_century

[53] Brundtland H. Our Common Future: World Commission on Environment and Development. UK: Oxford University Press; 1987

[54] Waters M, Plimmer F, Kenney S. Developer Strategies for Sustainable Development in the UK: Redevelopment Versus Refurbishment and the Sustainable Communities Plan. A Paper Presented at the Strategic Integration of Surveying Services, FIG Working Week. Vol. 13-17. Hong Kong SAR, China: International Federation of Surveyors FIG; 2007. pp. 1-15

[55] Department of the Environment, Transport and the Regions (DETR). Fuel Poverty: The New HEES. London: UK Department of Environment, Transport and the Regions; 1999

[56] ODPM. Sustainable Communities: Building for the Future. London: Office of the Deputy Prime Minister; 2003a

Chapter 4

Risks and Threats on the Coast of Quintana Roo: The Case of Tulum, México

Joel F. Audefroy

Abstract

The 160 km of the Riviera Maya from Cancun to Tulum are highly valued by investors in the tourism sector. In the case of the city of Tulum, the problem is more acute due to hydrometeorological risks and effects on the environment, such as mangroves and underground rivers. The development of basic urban infrastructure (water, sanitation, and solid waste) for the resident population does not allow the development of large tourism real estate projects that are the site of several controversies between environmentalists, landowners, the municipality itself and the investors. The existence of a Tulum National Park and the Sian Ka'an Biosphere Reserve put a brake on massive tourism development, but they are nevertheless affected. The methodology of analysis is based on a comparison between the risk maps and the municipal development plans, which reveals little interest in potential risks. Faced with this problem, the research proposes sustainable development and participatory risk management that will not affect the environment, including nature-friendly tourism development.

Keywords: risks, threats, urban development, tourism, investors

1. Introduction

The Municipality of Tulum was founded on March 13, 2008, from the territory of the municipality of solidaridad and is part of the Chichen-Itza-Coba-Tulum archaeological axis. It is located on the Mayan coast south of Cancun (132 km) and is part of the Cancun-Tulum tourist corridor. The geographical context of Tulum has long beaches and is situated in front of the largest chain of reefs in the world, wide extensions of mangroves, and surrounded by a jungle of great biodiversity, as well as numerous cenotes and caverns that are part of a system of underground rivers, in the northern and southern part of the urban area.

The city of Tulum is now facing pressure from the tourist real estate market, and its negative effects, such as real estate and urban speculation on communal land and environmental effects, such as sanitary landfills at the top of their capacity. There is also an environmental deterioration of hydrological systems, wetlands, coastal dunes, and coral reefs and insufficient infrastructure making the city vulnerable to all kinds of threats. In the year 2000, the Riviera Maya megaproject emerged, including Ciudad del Carmen and Tulum. The Rivera Maya was planned as a tourism megaproject for mass tourism without considering the effects, and it could have on the environment along the coast of Quintana Roo. The development of the City of Tulum should be limited by the Tulum National Park, founded in 1981, where the archaeological site (664 hectares) is located, and the Sian Ka'an Biosphere Reserve, both of which are considered natural protected areas (PNA). The Sian Ka'an Biosphere Reserve is part of the hydrological complex of lagoons, wetlands, underground rivers, and cenotes of the Mesoamerican Reef System and is located 10 kilometers south of Tulum and has an area of 528,147 hectares (ha). On January 20th, 1986, it was declared a biosphere reserve, as a World Heritage Site, by UNESCO. These wetlands form a natural barrier to weather disturbances, such as hurricanes, waves, and storm surges. The existence of these protected areas has done little to limit the expansion of tourist and housing developments in these areas. However, the areas that remain between the city itself and the coast are becoming real estate developments, such as Aldea Zama, despite occupying mangroves, gradually disappearing from the jungle.

The existence of ejido land around the city of Tulum, regulated by the urban development plan (2006–2030), predisposes future growth [1, 2]. These ejido lands are subject of high real estate speculation despite being jungle land. There are more than 14 tourist developments in Tulum National Park.

There is a conflict between the local authorities that approved an update to 2006 and urban development plan that proposed the construction of 60,000 hotel rooms within 25 years and the federal authorities demanding that the urban development plan conform to the legal framework without encroaching on federal powers. This urban development plan, updated in April 2008, was revoked in May 2008 by the supreme court of justice of the nation (SCJN) for invading the powers of the central state in the management and protection of natural protected areas and the Tulum National Park, thereby which the municipality of Tulum will not be able to assign population densities or assign land uses to the protected area, as it intended.

There are also conflicts between environmental groups, environmental NGOs, federal institutions and, on the other hand, local politicians, ejido owners, large international corporations, and private investors. The first conflict consists of a legal dispute over land ownership between the former owners and those requesting recognition of the expropriated territory for the national park and the archaeological zone.

There are several works about the environmental problems in Tulum and on the Mayan coast, most of them pose problems linked to tourism González [3–5], the coastal environment [6], urban resilience to hurricanes Camacho Sanabria José [7], effects of climate change on the coastal zones [8, 9], but few associate hydrometeorological risks with urban development caused by the growth of tourism.

The environmental problem manifests itself around four main axes:

- Hurricanes: Few are considered in the plans of tourism developers, but few are considered by the different factors as an important criterion.

- Flooding: Urban development plan (2006–2030) and the municipal development plan (2018–2021) do not consider the flood risk map considered in the municipal risk atlas [10]. The urban development plan was prepared before 2015 but does not justify this omission.

- The presence of underground rivers: South system: Sac Ox Belha and North system: Sac Actun, mentioned in the urban development plan were not considered by the plan itself in its urban growth plan, nor in the partial municipal plan.

- Sea level rise: Due to climate change, an effect of sea level rise between 0.5 and 1 m was estimated, the result of which shows a loss of 20% of the beaches and 90% respectively [8].

- The real estate invasion of the Sian Ka'an Biosphere Reserve.

These risks and threats are not contemplated in urban development plans, much less in city growth plans. It seems that the growth plans are unaware of these risks when they have been highlighted by various studies. The conceptual framework below explains in some way why these risks and threats have not been considered in the city's growth plans.

2. Conceptual framework

The conceptual framework is based on the following research questions: How are risks and vulnerability constructed in a coastal city? and How do risks and vulnerability grow over the years? To begin with, the importance of vulnerability is recognized, followed by risks and their social construction as central concepts in the occurrence of disasters that are by no means natural [11, 12]. Urban disasters are processes, sometimes long with multi-causal, social, economic, and political factors, particularly in cities like Tulum, which began to grow in the 1990s. This growth process illustrates how risk has been built over the last 30 years. Natural threats have not changed in this same period hurricanes, floods, permeable karstic soil, etc. What has changed is the occupation of the territory mainly due to pressure from the tourism industry. The social construction of risks has its origin in the creation of vulnerable conditions, for example, the construction of housing complexes on mangroves that thus disappear, and that are a natural barrier against the effect of hurricanes and limiting flooding.

Currently, new concepts are used by various researchers, such as adaptation and resilience. Although adaptation processes have always existed since the beginning of humanity, now it is about adaptation processes in the face of climate change that are promoted by international organizations. Likewise, several studies focus on "urban resilience", studying whether a city is resilient or not, these concepts may help to understand the phenomena but do not help to propose solutions to limit the vulnerability of cities to natural phenomena and socio-natural threats.

3. Methodology

An observation period of the municipality of Tulum goes between 1987 and 2020, corresponding to the data obtained in relation to urban and population growth between these dates. The analysis is developed from cartography prepared by the urban development plan (2006), the municipal development plan (2018-2021) and the municipal risk atlas (2015), and the partial program of urban development polygon south of the Tulum population center (2011). The proposal results from a comparison and superimposition layer of risk maps and urban growth and development maps carried out both by state and municipal authorities, as well as the private sector and tourism infrastructure promoters. Neither the municipal authorities nor the promoters of tourist infrastructures do this map crossing, if they had done so, their proposals in flood zones would be invalidated and unmarketable. For example, the flood zones marked in red in the risk atlas of Tulum do not appear in the urban development plans that have subdivided all the land around the urban center of Tulum. The mangrove areas located between the urban center and the coast have also been subdivided. Underground rivers no longer appear in the urban development plan. This method has aimed to highlight these omissions that leave the developer companies a free field for their projects without any restrictions other than those of the construction regulations.

4. Diagnosis: risks, threats, and vulnerability

Tulum suffers from two potential threats hydrometeorological events (hurricanes and floods) and anthropogenic threats caused by pressure from tourism and population growth. From a strictly environmental point of view, Tulum cannot grow due to the characteristics of its soil and subsoil (karstic soil, mangroves, and underground rivers). All growth brings with it conflicts between the environment and urbanization. The natural system is highly vulnerable: “Furthermore, the mangrove, like the reefs, is altered by the action of hydrometeorological phenomena, since during its action, salinity increases and exerting greater environmental pressure on this species” [13].

4.1 Hurricanes and floods

Since ancient times several hurricanes have impacted Quintana Roo and the Municipality of Tulum, we can observe that since 1951 eleven hurricanes have impacted with a category between three and five on the Saffir-Simpson scale, the coasts of Quintana Roo (see **Table 1**).

Hurricanes are generally accompanied by floods, strong winds and result in trees, and power poles falling. In Tulum, when a hurricane occurs, the internet is also cut off since the plant does not have a power generator, and the population is cut off from communication.

A study carried out by the University of Quintana Roo in 2019 [7] has shown through a survey, the percentages of the population of Tulum in relation to hurricanes to measure the level of urban resilience (**Table 2**).

This survey shows that more than 50% of the population that resides in Tulum is not prepared to face a hurricane of category two or more; however, 67% of the population has an emergency plan and knows the location of a temporary shelter. Most of the population is aware that their city is affected by hurricanes, but more than 50% have never experienced a hurricane and only 24% have received training on what to do in the event of a hurricane or flood.

Figure 1 clearly shows the upward trend from the year 2005 in affectations due to floods caused by hurricanes. In the municipality of Tulum, the flood zones are the following (**Figure 2**).

The entire coast where the main tourist infrastructures (hotels and bars) are located in a flood zone. In addition, the polygon considered for the extension of the

Name	Year	category	Most affected municipalities
Charlie	1951	4	Cozumel, Solidaridad
Janet	1955	5	Othon P. Blanco, Chetumal
Beulah	1967	5	Cozumel
Gilberto	1988	5	Cozumel, Solidaridad
Roxanne	1995	3	Felipe Carrillo Puerto, Tulum
Emily	2005	4	Tulum
Stan	2005	1	Chetumal
Wilma	2005	4	Cozumel, Solidaridad, Cancun
Dean	2007	5	Quintana Roo
Olga	2007	TS	Tulum
Dolly	2008	2	Puerto Morelos
Rina	2011	3	Tulum
Gamma	2020	1	Tulum
Delta	2020	4	Puerto Morelos, Tulum (efecto menor)
Zeta	2020	3	Tulum, Playa del Carmen

Source: Historical hurricane tracks (NOAA, 2021).

Table 1.
Hurricanes and tropical storms that impacted Quintana Roo.

Indicator	%
% of the population that has not experienced a hurricane	63
% of the population is unaware that the city is affected by hurricanes	7
% of the population that is aware that their home is located in a flood zone or that it may be affected by a hurricane	41
% of households that have a family emergency plan	67
% of homes insured against hurricanes and floods	7
% of households that have received training on what to do in the event of a hurricane or flood	24
% of the population that is aware of risk prevention programs for hurricanes and floods	32
% of the population that knows the location of the closest temporary shelter to their home	67
% of the population that knows the evacuation routes for hurricanes and floods	31

Source: [7].

Table 2.
Percentage of the population of Tulum in relation to their perception to be affected by hurricanes.

city in the Tulum Development Plan (2010) is also in a flood zone. This Polygon is considered an "urban reserve" for population growth (see **Figure 3**).

In summary, we have a population that is relatively unprepared to receive hurricanes and floods whose impacts are growing as of 2005, and a possibility of extension of the city of Tulum (Polygon South) in an area threatened by floods, and a coastal hotel zone as well as, subject to floods and hurricanes. We will see later that tourism megaprojects do not consider the threats clearly defined in the risk atlas as well as those defined in the urban development plan, nor the natural protected areas.

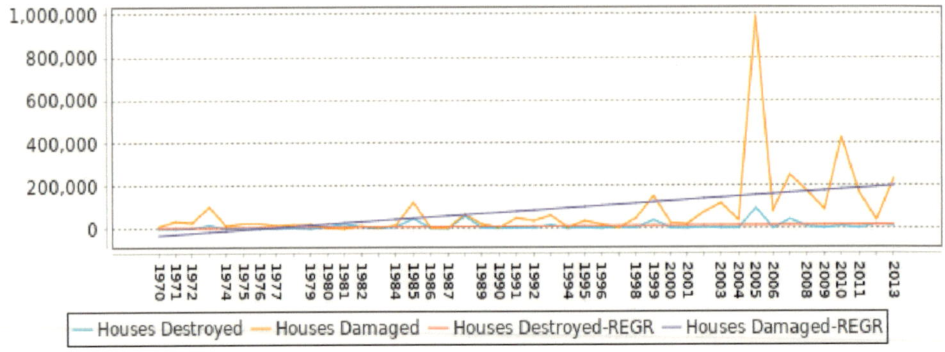

Figure 1.
Houses damaged and/or destroyed by floods in Quintana Roo in the period 1970–2013 (Source: Desinventar.net).

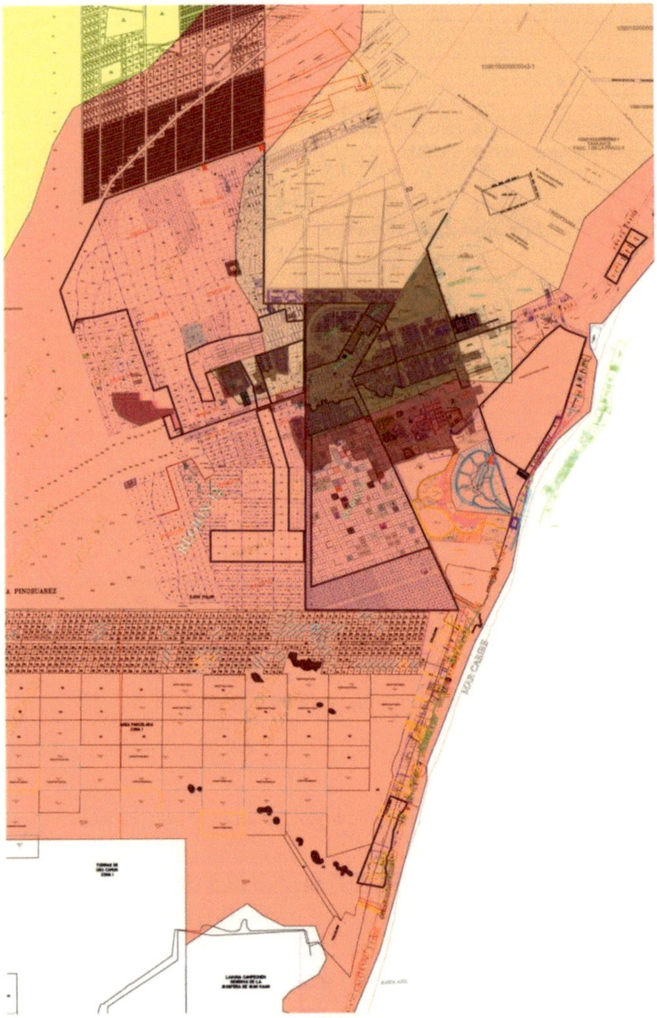

Figure 2.
Flood hazard map (in pink) (Sources: From the Risk Registrations of the Municipality of Tulum, 2015 and the Municipal Development Plan).

Figure 3.
Partial Development Program of the city of Tulum (Source: From the municipal development plan).

4.2 Water bodies and underground rivers

There are two underground river systems on each side of the city of Tulum: the southern system: Sac Ox Belha and the northern system: Sac Actun (map 3). These underground water body systems are connected to the cenotes. These systems are very vulnerable to all kinds of contamination, by sanitation systems: most houses have no drainage, only septic tanks whose absorption wells can contaminate underground rivers. The partial plan of the State of Quintana Roo, (2010) locates urban growth on the southern system of underground rivers (see map 4). The karstic rock soil is characterized by its high permeability; therefore, the infiltration of contaminants is very likely (**Figures 4** and 5).

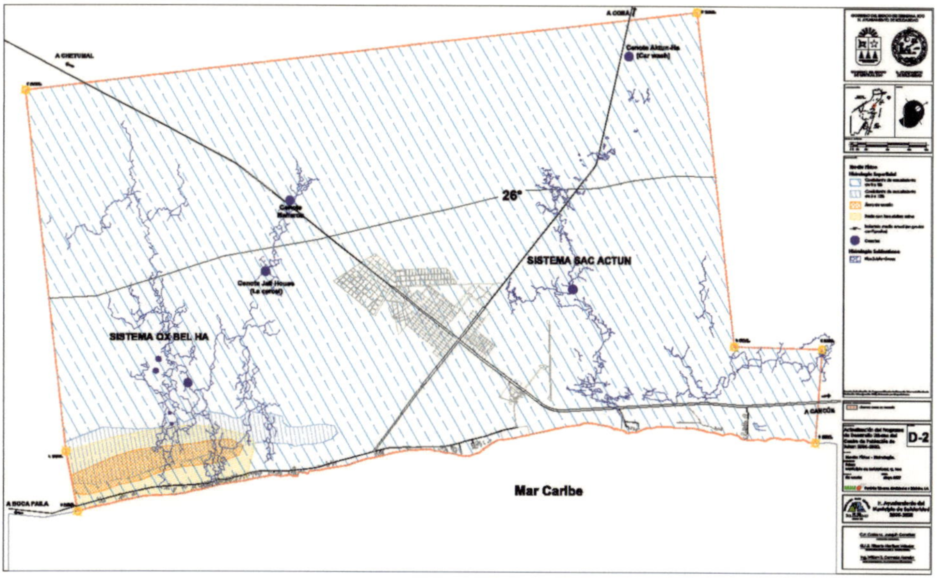

Figure 4.
Underground rivers: South system: Sac Ox Belha and North system: Sac Actun (Source: Urban Development Plan, 2006–2030).

Figure 5.
Growth area of the city of Tulum on the southern system of underground rivers (Source: from the Municipal Development Plan).

The current urban center does not have drainage and the porosity of the soil means that all matter is filtered and reaches the underground body of water. Fecal matter has already been found in the caverns. Urban growth without sewage treatment plants would lead to environmental and health catastrophes.

5. Solid waste management

According to data from the Tulum City Council, in 2014 around 22,636 m3 of (urban solid waste (USW) was collected and in 2015 around 25,606 m3 of municipal solid waste (MSW) was collected. It is estimated that an average of around 180 tons of USW is collected daily, which is deposited in an open-air dump that is located 9.5 km from the housing center on the road towards Coba. The dump is reaching its maximum capacity, to try to solve this problem the local authorities began, in 2010, steps to carry out a sanitary landfill that is located 18 km to the south of the city on federal highway 307 Chetumal-Cancun. The landfill came into operation in 2016 and receives between 100 and 120 tons/day of solid waste. However, no garbage separation system would allow the recycling of some materials, such as paper and cardboard, glass, and plastics. When the sargassum (*Sargassum fluitans and Sargassum natans*) arrives, in the hot season, there are no alternative places to process the brown algae, other than covering the algae with sand in places where there are no hotels. The growth of the population, as well as the growth of the floating population due to tourism, will increase the need for the collection and processing of MSW, and make the city more vulnerable to solid waste.

6. Interactions between urban growth, tourism, and hazards

A study carried out by the Universitat Rovira i Virgili, Spain, [5] based on the opinion on the impacts of climate change on the different actors involved in the management and promotion of the Riviera Maya (politicians, private sector, and associations) reveals three main impacts: sea level rise, major weather events (hurricanes), and coral bleaching. There is no mention of the threats of floods and droughts or the lack of basic services that could be disastrous in the event of a lack of drainage in the event of hurricanes. They consider that these three impacts are already generating negative effects on tourism activity. For the mitigation of these threats, most of the interviewees agree on the lack of public resources when the strong urban growth for tourism generates many resources for the municipalities of Tulum. The fragility or scarcity of basic services and the absence of mitigation measures have nothing to do with climate change, thus, these studies on the impacts of climate change make it possible to hide real problems and threats on the territory and contribute to building future disasters.

6.1 Map analysis

If we cross the flood map (from the risk atlas) with the urban growth plan (see map No. 2), we can see that a large part of the projected extension of the city is in a flood zone (very high in red). Most of the urban extension is intended for housing complexes for tourism, each lot or set of lots is labeled with the name of the owner. As for the underground river zones (see map No. 4) we can also observe that they are located below the growth zone of the city. In the medium or long term, it is evident to forecast that investments in tourism will be affected by floods caused by hurricanes or other meteorological phenomena. Likewise, the underground rivers will probably be more polluted by the black and gray waters of the housing complexes.

6.2 Evolution of marginalization in Tulum

The idea that the development of the tourist infrastructure could contribute to reducing the level of marginalization and the precariousness of the dwellings of the inhabitants is not evident if we compare the last three years. The percentage of the illiterate 15-year-old population grew due to the arrival of the migrant population looking for service and cleaning jobs. The percentage of occupants of dwellings without piped water decreased in 2015 but grew in 2020. Likewise, the number of occupants in

Years	2010	2015	2020
Total population	28263	32714	46 721
% Illiterate population aged 15 or over	8.26	5.97	29.67
% Occupants in private dwellings without drainage or toilet	9.94	3.10	3.64
% Occupants in private dwellings without piped water	6.5	0.95	2.85
% Occupants in private dwellings with dirt floors	5.38	4.19	7.43
% Private dwellings with overcrowding	55.58	43.83	43.24
% Employed population with incomes less than 2 minimum wages	30.91	23.39	51.26
Degree of marginalization	Low 0.76	Low 0.87	Low 0.88

Source: National Population Council (CONAPO).

Table 3.
Evolution of marginalization in Tulum from 2010 to 2020.

dwellings with dirt floors increased in 2020, only overcrowding decreased a little. In other words, there are no indications that the quality of life and housing has increased. The hypothesis stating that tourism development goes hand in hand with population and development cannot be verified in the case of Tulum (see **Table 3**).

7. Conclusions and recommendations

To reduce the impacts caused by the effects of climate and climate change, some basic principles are proposed. The city of Tulum is very vulnerable to these events and

Phenomena	Impacts	Basic Principles
Storm surges, storm surges, and hurricanes	Destruction of buildings, infrastructure, and natural systems. Erosion of beaches and coastal dunes. Flooding of buildings, infrastructure, open spaces, and natural systems.	1. Maintain the natural dynamics of floods, temporary outlets, storm surges, regular waves, and groundwater flows • Do not obstruct the flow of water • Do not obstruct sand deposition and dune formation, or their erosion • Maintain native vegetation in good condition 2. Build in the least exposed areas. 3. Maintain natural coastal protection systems (dunes, reefs, and mangroves).
Intense rains caused by north winds and storms	Flooding of buildings, infrastructure, open spaces, and natural systems. Soil erosion and the creation of gullies or ditches.	4. Maintain the natural dynamics of flooding, catchment, and release of the area • Maintain natural catchment and flooding areas • Create catchment areas and temporary flooding • Maintain or reconstruct the natural routes of water discharge towards areas of the flood. • Maintain or establish vegetation barriers to reduce erosion by runoff 5. Build elevated buildings above the flood elevation. 6. Disturb the slopes of the land and the vegetation as little as possible so as not to increase runoff.
Stronger and more frequent storm and hurricane winds	Destruction of buildings, infrastructure, and natural systems. Erosion of coastal dunes and beaches.	7. Maintain or restore the vegetation as a living barrier against the wind, according to the strength of the wind. 8. Build solid infrastructure to withstand winds. 9. Build open infrastructure to allow wind to pass through.
Sea level rise and waves	Erosion of beaches and coastal dunes Flooding of buildings, infrastructure, open spaces, and natural systems.	10. Maintain natural coastal protection systems: dunes, reefs, and mangroves. 11. Do not build structures that interrupt the coastal dynamics in dunes, lagoons, and wetlands 12. Maintain free areas where natural systems can migrate when increasing the sea level.

Source: From [9], Sobrevive al cambio climático.

Table 4.
Basic principles against climatic phenomena.

will be more so in the future, therefore it is important to know the coastal dynamics and not hinder natural processes through infrastructure or buildings. We summarize these basic principles in **Table 4**.

From these basic principles, recommendations can be made for land use practices and construction practices.

8. Land use practices

1. Areas adjoining the coastline: the construction of hotels or facilities should not be allowed; only light infrastructure should be allowed that does not impact the ground and that easily allows evacuation. In the municipality of Tulum, these are the protected natural areas (ANP).

2. In areas far from the coast, higher densities can be allowed (up to C.U.S. of 4.0) with a maximum of four levels but leaving a minimum of 50% of the land for green use and recharge.

3. A floor occupancy coefficient (C.O.S.) limit of 0.3 is recommended in coastal areas that contain tourist accommodation and up to 0.5 in urban areas.

4. Essential services, such as electrical power substations, hospitals, and health centers, should be located outside the storm surge zone of influence and outside the flood zone.

5. Do not obstruct the natural outlet areas towards the sea and keep them as green areas without construction or filling. Local knowledge and topographic surveys can determine these zones.

6. The buildings located interspersed allow the circulation of the prevailing winds.

7. Protect the mangroves because the roots and trunks cushion the impact of natural waves and storms on the beaches, thus reducing erosion, trapping debris, and generating soil.

9. Constructive practices and concepts

1. In flood-prone areas, it should be built on piles since it reduces the risk of flooding, facilitating the flow of tidal waves and reducing the temperature and humidity inside the room.

2. Use isolated triangular or oval footings to minimize the impact on the foundation due to the force of the water.

3. The open floor plan allows better ventilation inside the rooms and avoids the use of air conditioning.

4. The presence of an interior patio generates a different micro-climate from the exterior with native vegetation.

5. Preferably use heat-insulating materials, such as wood or bamboo in the construction of envelopes and roofs.

6. The portico allows shade and reduces sunlight on the walls, facing south or west.

7. Circular walls offer less resistance to hurricane winds (i.e. the traditional Mayan house).

8. Use lattice windows in case of large dimensions to better resist strong winds.

9. Cover with openings at the ends to ensure cross ventilation (chimney effect).

10. Obligation for all residential or tourist complexes to build a treatment plant for black and gray water.

11. Capture and store rainwater so as not to depend on the supply of drinking water, especially in areas with low-capacity aquifers, such as the fringes of dunes between the sea and coastal lagoons.

These recommendations consider the Tulum Urban Development Plan that foresees growth that will have a negative impact on the land, environment, transportation, and circulation. The way to limit the negative impacts is to follow these recommendations both at the level of land use and in construction practices. The demand for tourism infrastructure is such that large tourism companies have already bought properties and lots on the outskirts of the urban center of Tulum. In the next 10–15 years, the current Tulum will change into an important city in terms of population, which is why drastic measures are required from now on so that Tulum is a city that is resilient to such growth.

Acknowledgements

This research was financed by the Instituto Politecnico Nacional of Mexico, SIP N° 20210844. We appreciate the participation of Claudia Mitzy Jimenez Martinez, BEIFI student of the master's degree in the design of the maps.

Author details

Joel F. Audefroy
Instituto Politécnico Nacional (IPN), Escuela Superior de Ingeniería y Arquitectura, Tecamachalco (ESIA-TEC), México

*Address all correspondence to: takatitakite@gmail.com

References

[1] Gobierno Municipal de Solidaridad, Estado de Quintana Roo. Actualización del Programa de Desarrollo Urbano del Centro de Población de Tulum 2006-2030. México; 2007. p. 259

[2] Cesar MHJ, Enrique GS, Carlos MWR, Oscar FM. Gestión de desastres asociados a fenómenos hidrometeorológicos y climáticos en sistemas socio-ecológicos. México: CLAVE Editorial; 2021. p. 298

[3] Vera G, Angélica AC, Salvador CCBL. El desarrollo del turismo: ante el reto de una sociedad del riesgo y vulnerabilidad en la región costa maya, Quintana Roo, México. In: Bonnie Lucía Campos Cámara et al. Coord, editor. Sociedad del Riesgo en México: Análisis y perspectivas. Chetumal: Universidad de Quintana Roo; 2016. pp. 47-61

[4] Martinez R, Israel. Hacia una propuesta urbano ambiental de turismo sustentable, para la región costera de Tulúm. Roo, México, Tesis de Maestría, UNAM, México: Quntana; 2011. p. 172

[5] Santos-Lacueva Raquel y Saladié, Òscar. Cambio climático y vulnerabilidad del turismo en la Riviera Maya (México): ¿qué opinan los stakeholders?, Documento para su presentación en el VI Congreso Internacional en Gobierno, Administración y Políticas Públicas GIGAPP-IUIOG. Centro de Ciencias Humanas y Sociales. Consejo Superior de Investigaciones Científicas. (Madrid, España) del 29 Septiembre al 02 octubre de 2015. 2015

[6] la Lanza Espino Guadalupe D, Salvador H, Ivan P. Caracterización ambiental costera de la Riviera Maya, México, Diagnóstico costero de Quintana Roo. Saarbrücken, Alemania: Editorial Académica Española; 2012

[7] Manuel CSJ, Rosalía CA, David VT. Propuesta metodológica para medir la resiliencia urbana ante huracanes e inundaciones en el caribe mexicano. Revista de Estudios Latinoamericanos sobre Reducción del Riesgo de Desastres (REDER), Santiago de Chile. 2009;**3**(2):28-43

[8] Márquez García AZ, Márquez García E, Recaséns ABC, Torres Rodríguez V. Cambio en la línea de costa en la Rivera Maya debido a fenómenos hidrometeorológico extremos, ¿Consecuencia del cambio global cli- mático? In: Botello AV, Villanueva-Fragoso S, Gutiérrez J, Rojas Galaviz JL, editors. Vulnerabilidad de las zonas costeras mexicanas ante el cambio climático. Campeche: Universidad Autónoma de Campeche; 2010. pp. 345, 514-358

[9] Silva H, Rosas G, Secaira F, Meller T, Mendoza M. Sobrevive al cambio climático. Catálogo de buenas prácticas para reducir los impactos del clima en la zona costera de Quintana Roo, México. Chetumal, Quintana Roo: Instituto Tecnológico de Chetumal, The Nature Conservancy, Amigos de Sian Ka'an-marti; 2014

[10] SEDATU. Atlas de riesgos naturales del Municipio de Tulum Quintana Roo. México: SEDATU; 2015. p. 198

[11] García-Acosta V. La Antropología de los Desastres en América Latina: Estado del arte. México: Centro de Investigaciones y Estudios Superiores en Antropología Social; 2021

[12] Oliver-Smith A. Hazards and disasters research in contemporary anthropology. In: Wright JD, editor. International Encyclopedia of the social and Behavioral Sciences. Amsterdam: Elsevier; 2015. pp. 546-553

[13] Ayuntamiento de Tulum. Plan Municipal de Desarrollo de Tulum 2018-2021, México, 137 p. 2018

Chapter 5

Evaluation of Net Radiation in San Luis Potosí City – México, with Remote Sensing Processes

Abraham Cárdenas-Tristán, Andrés Gerardo Castro-Ovalle, Oscar Reyes-Cárdenas and María Guadalupe Galindo Mendoza

Abstract

Net radiation is essential in analyzing the earth's energy balance since it regulates the soil's temperature, humidity, and thermal processes. Then, energy regulates an urban area's climate and heat distribution. Therefore, problems such as heat islands arise if these areas have more impervious surfaces than vegetation cover. The main objective of this study is to evaluate the net radiation in the city of San Luis Potosí from a multi-temporal analysis applying remote sensing. Satellite images from the Landsat 5 TM and 8 OLI/TIRS sensors were used. Through remote sensing processes, the imagery has been calibrated to obtain the net flux extracting the incident and outgoing radiation. Our results show that net radiation levels decrease in the dry season and increase in the rainy season. It was also observed that the incident radiation predominates over the outgoing radiation on the city's impervious surfaces, causing temperature increase. This study is helpful in decision-making related to the city's planning.

Keywords: net radiation, remote sensing, multitemporal analysis, urban climate, Landsat imagery

1. Introduction

All bodies in the universe with a temperature greater than zero can emit radiation. After the emission of radiation, the next step is its propagation to a second body through the air or space, where different processes arise: reflection, scattering, absorption, refraction, dispersion, and transmission of radiation [1]. Solar radiation is a term that refers to the energy flow produced by the sun in the form of photons or electromagnetic waves [2], which provides energy to the atmosphere, clouds, and the earth's surface. How energy flows to enter and leave the earth's surface is known as the surface energy balance. This process determines the surface temperature and is closely related to the atmosphere and surface elements (e.g. water, rocks, and vegetation) [3].

Net radiation is a crucial factor in the energy balance of the surface, serving to study characteristics of the earth's surface such as albedo, emissivity, temperature, and humidity, and thermal processes of the soil. Also, it is the leading actor in processes such as soil and air heating, photosynthesis, evapotranspiration, crops and water resources [4–6]. In an urban environment, net flux regulates climate and heat distribution. Therefore, when these areas grow uncontrollably and impermeable materials replace the earth's surface, the energy balance of the surface is affected, and problems such as heat islands arise [7, 8]. Besides, the heterogeneous nature of urban areas causes the kinetic and thermodynamic properties of the surface to change dramatically if the physical characteristics of the soil are modified. As a result, this causes cities to have unique microclimates [6].

Generally, net radiation is measured from meteorological stations with the help of radiometers. However, these expensive instruments do not effectively represent the net flux variation throughout the territory and do not accurately reflect its effects on the earth's surface [6]. However, other techniques have been developed to measure net radiation effectively. Remote sensing is a science focused on acquiring the electromagnetic energy emitted or reflected by the earth in the form of digital images to obtain information about an object without being in direct contact with it [9]. In addition, data can be continuously produced with high spatial resolutions and at a regional scale [10]. Net radiation retrieval through remote sensing allows its estimation in topographically complex areas. Radiative transfer models, meteorological data, and Digital Elevation Models (DEM) are often necessary to compute net flux [10, 11]. Various technological advances have been made these days by applying remote sensing to estimate net flux with high spatial resolution and precision, trying to reduce dependence on meteorological data [7, 10–14].

The city of San Luis Potosí has experienced gradual and unplanned growth during the last decades, mainly in the city's peripheral areas [15], which in conjunction with the lack of green spaces in the area, has caused an imbalance in the climate, raising temperatures in some parts of the city. A study carried out in the city of San Luis Potosí shows that deforestation worsens the heat island phenomenon; however, as it is only a descriptive methodology of the urban tree canopy [16], it is not thoroughly analyzed why this occurs. Therefore, it is essential to understand how net radiation behaves in the city and what factors affect the urban climate. This work's main objective is to evaluate net radiation behavior in the city of San Luis Potosí through a multi-temporal analysis applying remote sensing. This analysis will serve to determine the net flux influence degree on the urban climate of the area, which is crucial to support decision-making for the improvement of urban planning and the establishment of mitigation measures for heat island phenomena.

2. Remote sensing processes for net radiation computation

2.1 Study area

The city of San Luis Potosí is the capital of the San Luis Potosí State, which is part of the 32 states that make up the Mexican Republic, located in the north-central zone of the state between the parallels 22.2214° and 22.0058° North latitude and the meridians −101.0566° and −100.7743° West longitude (**Figure 1**). According to the 2020 Population and Housing Census derived from the National Institute of Statistics and Geography (INEGI) [17], the study area has a population of 1,243,980. The city has an average elevation of 1860 meters above mean sea level and is considered an area with a dry-semi-desert climate. The average level of precipitation is 245 mm. However, this value varies depending on the year. The mean temperature is 16.8°C.

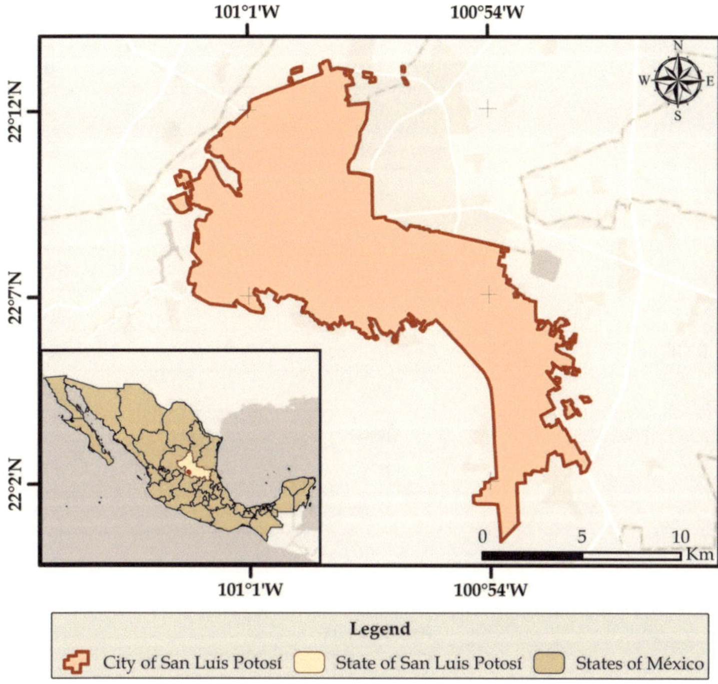

Figure 1.
Study area location.

2.2 Data acquisition

In order to evaluate energy throughout the San Luis Potosí city, it was necessary to obtain a DEM at a resolution of 30 m from INEGI. Also, satellite images from the Landsat 5 TM and 8 OLI/TIRS sensors were used, with a spatial resolution of 30 m.

The study was developed for the period 1990–2017, and two representative images of the region's rainy and dry seasons per year were needed. **Table 1** describes the characteristics of the sensors used. Landsat 5 TM can record the electromagnetic energy reflected and emitted by the earth's surface in seven different wavelengths (three bands in the visible spectrum, one near infrared, one shortwave infrared and one thermal infrared). While Landsat 8 has two instruments: OLI, composed of nine bands from the visible spectrum to the short-wave infrared, and the TIRS sensor, consisting of two spectral bands in the thermal infrared region [18].

Table 2 shows the images used by each sensor for path and row 28/45. These images were downloaded from the Earth Explorer page (https://earthexplorer.usgs.gov/) managed by the US Geological Survey (USGS). For the Landsat 5 TM sensor, ten images acquired in the period 1990–2009 were used, and for Landsat 8, four images were downloaded in the period 2015–2017.

2.3 Preprocessing for Landsat imagery

The digital number of the satellite images must have a standard and representative scale to calculate the net flux. For this reason, it is necessary to apply a radiometric

Sensor	Spectral band	Wavelength (μm)
Landsat 5 TM	Blue	0.45–0.52
	Green	0.52–0.60
	Red	0.63–0.69
	NIR	0.76–0.90
	SWIR 1	1.55–1.75
	Thermal	10.40–12.50
	Swir 2	2.08–2.35
Landsat 8 OLI/TIRS	Costal/Aerosol	0.432–0.451
	Blue	0.452–0.512
	Green	0.533–0.590
	Red	0.636–0.673
	NIR	0.851–0.879
	SWIR 1	1.566–1.651
	SWIR 2	2.107–2.294
	Panchromatic	0.503–0.676
	Cirrus	1.363–1.384
	Thermal 1	10.60–11.19
	Thermal 2	11.50–12.51

Table 1.
Landsat 5 TM and Landsat 8 OLI/TIRS spectral bands.

correction. In this case, the methodology proposed by Chander et al. [19] (Eq. (1)) was used. The process described was developed for the Landsat 5 TM sensor; however, it can also be used for the Landsat 8 sensor, considering that the gain and bias factors are band-specific values and can be found in the image metadata.

$$L_\lambda = G_{rescale} \times Q_{cal} + B_{rescale} \quad (1)$$

where L_λ is the spectral radiance (W/m^2 sr μm); Q_{cal} is the digital number of the pixel; and $G_{rescale}$ and $B_{rescale}$ are band-specific rescaling factors of gains and bias from the sensor. For Landsat 8 OLI/TIRS $G_{rescale}$ is RADIANCE_MULT_BAND_x and $B_{rescale}$ is RADIANCE_ADD_BAND_x from metadata.

2.4 Net radiation calculation

For this study, energy retrieval was computed using the processes described by Allen et al. [20]. Net flux is the energy flow from the sun and the earth's surface [14]. It is computed by extracting the incoming short wave and long-wave radiation and the outgoing long-wave radiation (Eq. (2)):

$$Rn = Rs\downarrow - \alpha Rs\downarrow + RL\downarrow - RL\uparrow - (1 - \varepsilon_0)\, RL\downarrow \quad (2)$$

Acquisition date	Season	Sensor
January 27, 1990	Dry	Landsat 5 TM
May 3, 1990	Rainy	Landsat 5 TM
February 10, 1995	Dry	Landsat 5 TM
October 8,1995	Rainy	Landsat 5 TM
January 23, 2000	Dry	Landsat 5 TM
August 18, 2000	Rainy	Landsat 5 TM
April 26, 2005	Dry	Landsat 5 TM
July 31, 2005	Rainy	Landsat 5 TM
March 4, 2009	Dry	Landsat 5 TM
July 26, 2009	Rainy	Landsat 5 TM
January 16, 2015	Dry	Landsat 8 OLI/TIRS
July 27, 2015	Rainy	Landsat 8 OLI/TIRS
February 22, 2017	Dry	Landsat 8 OLI/TIRS
October 20, 2017	Rainy	Landsat 8 OLI/TIRS

Table 2.
Satellite images used to obtain net radiation.

where $Rs\downarrow$ is the incident shortwave radiation (W m^{-2}); α is the superficial (dimensionless) albedo; $RL\downarrow$ is the incident long-wave radiation (W m^{-2}); $RL\uparrow$ is the reflected long-wave radiation (W m^{-2}); and ε_0 is broad band (dimensionless) surface thermal emissivity. The term $(1 - \varepsilon_0)\ RL\downarrow$ represents the incident long wave fraction reflected from the Earth's surface.

2.4.1 Incident wave radiation

Defined as the difference between the incident short-wave radiation (W m^{-2}) on the Earth's surface and the short-wave radiation reflected to the atmosphere (Eq. (3)): $R_{s\downarrow}$ is one of the essential fluxes to estimate the net radiation, its calculation was carried out from a Digital Elevation Model of the study area:

$$R_{s\downarrow} = \frac{G_{SC} \cos \theta_{rel} \tau_{s\omega}}{d^2} \tag{3}$$

where G_{SC} is the constant of solar (1367 W m^{-2}), θ_{rel} is the solar incidence angle, d^2 is the relative distance between the earth and the sun (Eq. (9)), and $\tau_{s\omega}$ is transmissivity of the atmosphere, to determine the level of atmospheric transmissivity present in the study area, we use the function expressed in the Eq. (4). Knowing the transmissivity of the atmosphere is crucial to determine the transmissivity $R_{s\downarrow}$ value. $\tau_{s\omega}$ for the study area was determined from its atmospheric pressure and the water content in the atmosphere. (Eq. (4)).

$$\tau_{s\omega} = 0.35 + 0.627\ exp\left[\frac{-0.00146\,P}{K_t\cos\theta_{hor}} - 0.75\left(\frac{W}{\cos\theta_{hor}}\right)^{0.4}\right] \tag{4}$$

where P is the atmospheric pressure (kPa), W is the water content in the atmosphere (mm), and θ_{hor} is the zenithal solar angle on a horizontal surface. K_t corresponds to a turbidity coefficient of $0 < K_t < 1.0$, in which $K_t = 1.0$ is used in areas with clean air, while $K_t = 0.5$ is used in areas with extreme turbidity, dust, or polluted air. P calculating its variation with respect to elevation pixel by pixel using a Digital Elevation Model (DEM) and applying the following formula (Eq. (5)):

$$P = 101.3\left(\frac{293 - 0.0065\,z}{293}\right)^{5.26} \tag{5}$$

where 293 is the standard air temperature (K) and z is the altitude above sea level (m). W (mm) is obtained from near-surface vapor pressure data, which are obtained from meteorological stations within the study area by applying (Eq. (6)). In this work, e represents vapors pressure near the surface and was calculated pixel by pixel from the dew point recorded in the study area. This represents an advantage because W was not only taken as a constant value of a meteorological station but its special variation throughout the study area was determined.

$$W = 0.14e_aP_{air} \tag{6}$$

The solar incidence angle is defined as the angle existing between the solar beam and a vertical line perpendicular to the earth's surface. In order to calculate θ_{rel}, slope and aspect are extracted from the DEM and Eq. (7) is applied:

$$\begin{aligned}\cos\theta_{rel} = {} & \sin\delta\sin\varphi\cos s - \sin\delta\cos\varphi\sin s\cos\gamma + \cos\delta\cos\varphi\cos s\cos\omega \\ & + \cos\delta\sin\varphi\sin s\cos\gamma\cos\omega + \cos\delta\sin\gamma\sin s\sin\omega\end{aligned} \tag{7}$$

where δ is the Earth' declination (positive during summer in the northern hemisphere); φ is the pixel latitude; s is the slope; γ is the surface aspect angle; and ω is the hour angle.

2.4.2 Surface albedo

Surface albedo is the quantity of radiation that is reflected by the earth's surface; it is a biophysical characteristic that regulates net flux. Its value varies according to the land cover, and in some cases, a low albedo value (as it is in the case of urban cover) represents an increase in temperature. It is estimated by using wavelengths capable of reflecting energy (Eq. (8)).

$$a_s = \sum_{b=1}^{n}\left[\rho_{s,b}w_b\right] \tag{8}$$

where W_b is the weighting coefficient representing the solar radiation fraction on the surface that occurs within the spectral range in a determined band [20]; n is the number of bands of the sensor; and $\rho_{s,b}$ is the surface reflectance of each band, which was calculated using the Fast Line-of-sight Atmospheric Analysis of Spectral Hypercubes (FLAASH) method, eliminating aerosols' effect, intrinsic radiance introduced into the sensor and reflected in the image. This method includes the MODTRAN radiative transfer code [21].

2.4.3 Reflected long-wave radiation

This flux is a surface-driven radiation produced by temperature and emissivity, and to obtain it we used the Stefan-Boltzmann equation (Eq. (9)): The calculation of this variable was used to understand how the land surface of the study area is heated.

$$R_{L\uparrow} = e_0 \sigma {T_S}^4 \tag{9}$$

where e_0 is the surface emissivity (dimensionless); σ is Stefan Boltzmann constant ($5.67 \times 10–8$ W m^{-2} K^{-4}); and T_S is the temperature of surface (K). The surface emissivity is estimated using the Eq. (10). e_0 was obtained from the LAI (Leaf Area Index), which measured the relationship between the total leaf area per unit of land area. e_0 allowed us to know the degree of relationship between land cover in the study area and its temperature.

$$e_0 = 0.95 + 0.01\, LAI\ when\ LAI \leq 3; e_0 = 0.98\ when\ LAI > 3 \tag{10}$$

Surface temperature is a direct indicator of the earth's surface temperature, which can vary due to geological, geophysical, and geochemical parameters of the Earth and atmosphere [22, 23]. This parameter can be calculated by applying the modified Plank equation, which includes atmospheric corrections and surface emissivity (Eq. (11)):

$$T_S = \frac{K_2}{\ln\left[\left(\varepsilon_{NB}\frac{K_1}{R_c}\right) + 1\right]} \tag{11}$$

where ε_{NB} corresponds to the short-wave emissivity of the satellite thermal sensor and R_c is the fixed thermal radiance of the surface, which is obtained from the spectral radiance of the thermal bands (Eq. (12)). For Landsat 5 TM, K_1 = 607.8 and K_2 = 1261 W m^{-2} sr^{-1} μm^{-1}. For thermal 1 band of Landsat 8, constants K_1 = 774.89 and K_2 = 1321.08 W m^{-2} sr^{-1} μm^{-1}.

$$R_c = \frac{L_{t,T} - R_p}{\tau_{NB}} - (1 - \varepsilon_{NB})R_{sky} \tag{12}$$

where $L_{t,T}$ is the spectral radiance of the thermal bands (Wm^{-2} sr^{-1} μm^{-1}); R_p is the radiance path at length 10.4–12.5 μm Wm^{-2} sr^{-1} μm^{-1}; R_{sky} is the short-wave thermal radiation in clear sky conditions; and τ_{NB} is the transmissivity of air in the length range of 10.4–12.5 μm. The Atmospheric Correction Parameter Calculator [24] was used to calculate R_p, τ_{NB} and R_{sky}, which are values that require an atmospheric correction model. These parameters were calculated exclusively for each representative image of the study area, considering its characteristics. These parameters are the following: date of image acquisition, location, atmospheric model of the study area and terrestrial conditions (elevation, average temperature, and atmospheric pressure).

Surface emissivity for the thermal band is calculated through the narrow band transmissivity, and for this work was calculated using band 6 of sensor 5TM and band 10 of sensor 8 TIRS. (Eq. (13)):

$$\varepsilon_{NB} = 0.97 + 0.0033\, LAI, for\ LAI \leq 3; \varepsilon_{NB} = 0.8, for\ LAI > 3 \tag{13}$$

Eq. (13) is applied when the Normalized Difference Vegetation Index (NDVI) is greater than zero. NDVI less than zero indicates the presence of water, snow or ice, resulting in $\varepsilon_{NB} = 0.985$.

2.4.4 Incident long-wave radiation

The incoming long-wave radiation is an atmospheric flux from the thermal region of the electromagnetic spectrum (W m^{-2}) and is calculated by the Stefan-Boltzmann equation (Eq. (14)): $R_{L\downarrow}$ depends on atmospheric emissivity and surface temperature, so it is important that this flux be determined according to the characteristics of the study area.

$$R_{L\downarrow} = \varepsilon_a \sigma T_a{}^4 \tag{14}$$

where ε_a is the effective atmosphere emissivity (dimensionless) and $T_?$ is the air temperature near the surface (K). ε_a calculation can be performed using Eq. (15): ε_a depends on $\tau_{s\omega}$, therefore in the studio area $\tau_{s\omega}$ was calculated pixel by pixel to record its spatial variability, allowing ε_a to not just be taken as a constant value.

$$\varepsilon_a = 0.85(-\ln \tau_{s\omega})^{0.09} \tag{15}$$

where $\tau_{s\omega}$ is the broadband atmospheric transmissivity of short-wave radiation derived from Eq. (4). The parameter T_a can be replaced by surface temperature (T_s).

3. Net radiation assessment in the city of San Luis Potosí

3.1 Dry season

The net radiation in the dry season in the city of San Luis Potosí was calculated from a representative month (January to May, November and December). **Figure 2** shows the maps with the energy values for 1990–2017. In 1990 there was more incident radiation than reflected radiation, with a maximum value of 695.3 W m^{-2} and a minimum value of −61.3 W m^{-2}. For this year, most of the study area tends to have high net flux values because, in 1990, the city was not so developed, and there was a more significant presence of vegetation, unlike impermeable materials. In 1995 the maximum value of energy decreased to 523.3 W m^{-2}. These values are mainly found in urban coverage, where materials such as pavements and roofing absorb more incident energy. While the minimum radiation values are in the vegetation zones, some of these areas present negative values, which indicate that an energy deficit is being experienced (a slight cooling of the surface).

Further, in 2000 and 2005, it is observed that the city presents average values between 80 and 60 W m^{-2}. The lowest values are in areas with vegetation, indicating that most of them are outgoing radiation. In 2009, the highest net flux was

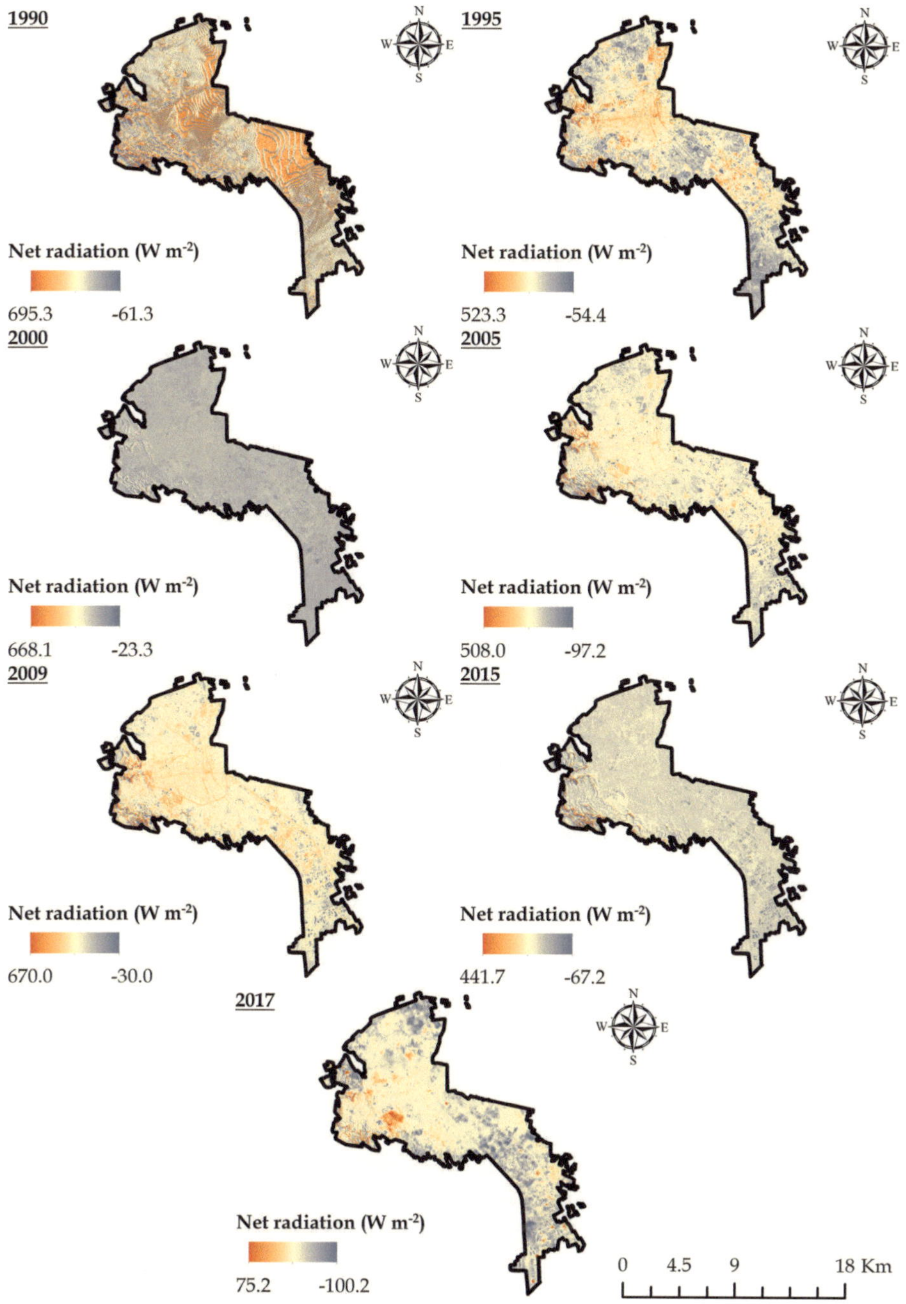

Figure 2.
Maps of net radiation in the dry season for 1990-2017.

concentrated in mountainous regions. Values of approximately 100 W m^{-2} can be observed on the city's impervious surfaces, and some minimum values (-30 W m^{-2}) were in farmland areas. In 2015, net flux below 100 W m^{-2} was obtained. In 2017 the behavior was very peculiar because the energy recorded was very low, which means there was less incident radiation concerning the outgoing, resulting in a warming of the study area.

3.2 Rainy season

The energy estimation in the rainy season was performed considering a representative month (June to October) as a sample. **Figure 3** shows the net flux behavior in the rainy season from 1990 to 2017. In 1990 the highest net flux (322.1 W m^{-2}) was in the impervious covers and the lowest (around −84 W m^{-2}) was in vegetation covers. The behavior is very similar to the dry season, although the level of net flux is lower. In 1995, the incident radiation exceeded outgoing radiation. Values of 693.0 W m^{-2} were observed on surfaces such as pavements and roofing. In contrast, values close to −57.4 W m^{-2} were found in vegetation areas. In 2000, most of the city presented energy of approximately 200 W m^{-2}, relief zones had values close to 625.6 W m^{-2}, and negative values (−14.9 W m^{-2}) continued to be concentrated in vegetation areas. However, for this year, negative values were lower.

On the other hand, the maps show similar behavior for the years 2005 and 2009. The net flux values are between −94 and 580 W m^{-2}. The highest net flux was in some of the city's main avenues and the lowest was in farmland and vegetation areas. In 2015 it was observed that the energy deficit was more significant than in 1995 in the dry season since most of the city's territory registered values close to −100 W m^{-2}, which suggests that there was a cooling in surface temperature because the outgoing radiation exceeded the incident radiation. In 2017, the city mainly presented energy close to 30 W m^{-2}. During the period 1990–2017, it was observed that generally, in the rainy season, the net flux increases concerning the dry season, which is typical behavior in summer [10]. Also, the results show that in the city's impervious surfaces, the incident radiation is greater than the outgoing radiation, which means that in these areas, the energy cannot be used by processes such as evapotranspiration, causing an increase in temperatures in the city.

4. Net radiation changes implications

During the period 1990–2017, it was observed that, in the rainy season, net flux increases compared to the dry season; this is considered a typical behavior in summer [10]. Related to the surface temperature values recorded in the area, it has been found that in the dry seasons, the surface temperature increases, mainly between March to May [25].

On the other hand, in this study, it was found that the city's impermeable materials have the highest values of net flux, meaning that, in these areas, the energy cannot be used for the evapotranspiration processes, causing an increase in the temperature of the city. According to Ovalle et al. [25], it has been verified that the surfaces with coverage without vegetation and urban areas tend to have higher temperatures in the study area. Likewise, the study area has presented climatic changes related to soil cover [25, 26]. Several analyses registered similar behaviors [27–29], where the energy depends on the ground cover. Impermeable surfaces such as asphalt mainly store the incident radiation and reflect less radiation, producing increases or decreases in surface temperature and causing phenomena such as urban heat islands.

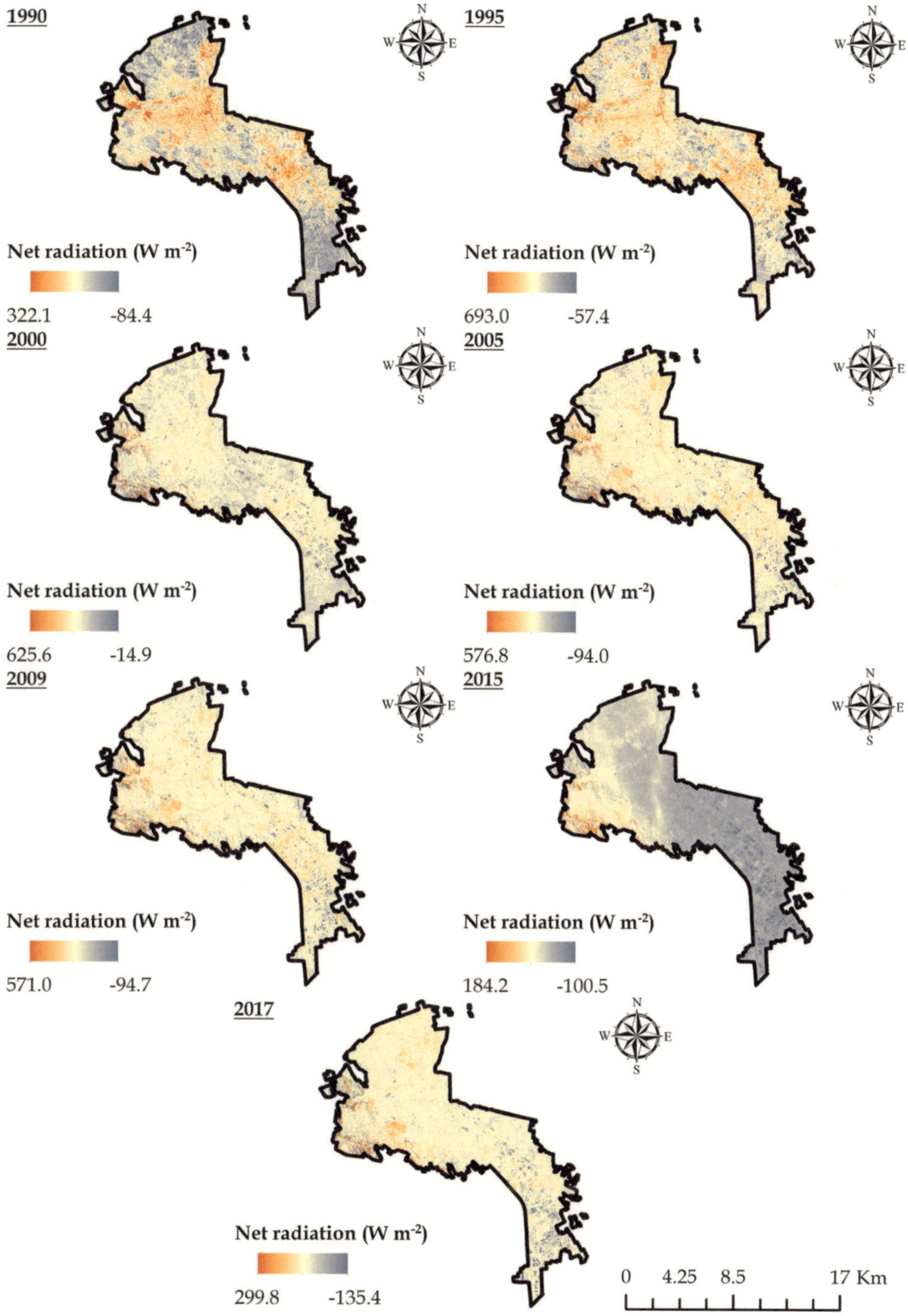

Figure 3.
Maps of net radiation in the rainy season for 1990-2017.

In this study, images from the Landsat 5 TM and 8 OLI/TIRS satellites were used because they have a sensor capable of recording wavelengths corresponding to thermal infrared, which is essential in the methodology used to calculate the net radiation. However, the methodology could be adapted to Sentinel 2 sensors since its images have been found to have greater spatial detail and help estimate net flux in regions with complex terrain. Unlike Landsat sensors, which, due to their temporal resolution, do not allow the analysis of detailed canopy information [30]. Also, the energy analysis in San Luis Potosí city could be complemented with land cover maps to have greater certainty of the effects caused by net radiation.

5. Conclusions

In this study, an evaluation of the energy in San Luis Potosí city was performed. This multi-temporal analysis is essential to know the net flux behavior throughout the study area and how it affects its climate. In addition, it is a turning point in the energy flow exploration of the soil in the San Luis Potosí city. This analysis can support decision-making on how the city's development is being planned and if it is considering how the urban climate interferes with the population's comfort.

The climate in an urban environment is a multifactorial problem since its variation is due to different causes; the natural soil replaced by impervious surfaces, the reduction of vegetation or the release of greenhouse gases. Therefore, one way to study the climate of a city is through its energy balance. Net radiation is a crucial factor in studying the earth's surface characteristics, such as albedo, emissivity, temperature and humidity and processes. According to the results obtained in this work, the San Luis Potosí city, in the dry season, has low levels of net flux. In the rainy season, there is higher energy. This behavior is typical for winter and summer. However, in both seasons, it was observed that incident radiation predominates over outgoing radiation, which means a problem for the city, which has a significant presence of impervious materials. These surfaces retain energy and emit it in the form of an increased temperature-producing urban heat islands.

This paper proposes to analyze the urban climate of the San Luis Potosí city through net radiation. The methodology includes a multi-temporal analysis to observe the net flux behavior related to its unplanned growth. The method used to estimate the energy was adequate for the study area to produce high-precision maps. Also, the net flux calculation through remote sensing processes was of great help since indicators with a high spatial resolution were generated, showing the net radiation's variability through space and time. However, for future research, determining the city's energy balance and including information on land cover in the study area could improve the urban climate analysis.

Acknowledgements

The authors of this work thank the United States Geological Survey (USGS) for making this work possible by processing and distributing the satellite images and Centro de Investigación y Estudios de Posgrado, Facultad de Ingeniería, Universidad Autónoma de San Luis Potosí (UASLP) and the Programa educativo de Ingeniería en Geomática. We are also grateful for the cooperation of CIACYT—Laboratorio Nacional de Geoprocesamiento de Información Fitosanitaria (LaNGIF).

Conflict of interest

The authors declare no conflict of interest.

Author details

Abraham Cárdenas-Tristán[1*], Andrés Gerardo Castro-Ovalle[2], Oscar Reyes-Cárdenas[1] and María Guadalupe Galindo Mendoza[3]

1 Centro de Investigación y Estudios de Posgrado, Facultad de Ingeniería, Universidad Autónoma de San Luis Potosí, San Luis Potosí, Mexico

2 Facultad de Ingeniería, Universidad Autónoma de San Luis Potosí, San Luis Potosí, Mexico

3 CIACYT—Laboratorio Nacional de Geoprocesamiento de Información Fitosanitaria (LaNGIF), Facultad de Ingeniería, Universidad Autónoma de San Luis Potosí (UASLP), San Luis Potosí, Mexico

*Address all correspondence to: abraham.cardenas@uaslp.mx

References

[1] Jacobson MZ. Fundamentals of Atmospheric Modeling. 2nd ed. Cambridge: Cambridge University Press; 2005. 829 p. DOI: 10.1017/CBO9781139165389

[2] Duro-Mota E, Campillos-Páez MT, Causín-Serrano S. El sol y los filtros solares. Medifam [Internet]. 2003;13(3). Available from: http://scielo.isciii.es/scielo.php?script=sci_arttext&pid=S1131-57682003000300005&lng=en&nrm=iso&tlng=en

[3] Van den Broeke M, Fettweis X, Mölg T. Surface energy balance. In: Singh VP, Singh P, Haritashya UK, editors. Encyclopedia of Snow, Ice and Glaciers. Dordrecht: Springer Netherlands; 2011. pp. 1112-1123. DOI: 10.1007/978-90-481-2642-2

[4] Geraldo-Ferreira A, Soria-Olivas E, Gómez-Sanchis J, Serrano-López AJ, Velázquez-Blazquez A, López-Baeza E. Modelling net radiation at surface using "in situ" netpyrradiometer measurements with artificial neural networks. Expert Systems with Applications. 2011;**38**(11):14190-14195. DOI: 10.1016/j.eswa.2011.04.231

[5] Ayoola MA, Sunmonu LA, Bashiru MI, Jegede OO. Measurements of net all-wave radiation at a tropical location, Ile-Ife, Nigeria. Atmósfera. 2014;**27**(3):305-315. DOI: 10.1016/S0187-6236(14)71118-X

[6] García-Cueto R, Santillán-Soto N, Haro-Rincón Z, Ojeda-Benítez S, Bojórquez-Morales G. Parameterization of net radiation in an arid city of northwestern Mexico. Atm. 2015;**28**(2): 71-82. DOI: 10.20937/ATM.2015.28.02.01

[7] Chakraborty SD, Kant Y, Mitra D. Assessment of land surface temperature and heat fluxes over Delhi using remote sensing data. Journal of Environmental Management. 2015;**148**:143-152. DOI: 10.1016/j.jenvman.2013.11.034

[8] Kuang W, Li Z, Hamdi R. Comparison of surface radiation and turbulent heat fluxes in Olympic Forest Park and on a building roof in Beijing, China. Urban Climate. 2020;**31**:100562. DOI: 10.1016/j.uclim.2019.100562

[9] Lillesand TM, Kiefer RW, Chipman JW. Remote Sensing and Image Interpretation. 7th ed. Hoboken, NJ: John Wiley & Sons, Inc; 2015. p. 720

[10] Liu X, Zhang J, Yan H, Yang H. Estimation of the surface net radiation under clear-sky conditions in areas with complex terrain: A case study in Haihe River basin. Frontiers in Ecology and Evolution. 2022;**10**:935250. DOI: 10.3389/fevo.2022.935250

[11] Duguay CR. An approach to the estimation of surface net radiation in mountain areas using remote sensing and digital terrain data. Theoretical and Applied Climatology. 1995;**52**(1–2): 55-68. DOI: 10.1007/BF00865507

[12] Hu D, Cao S, Chen S, Deng L, Feng N. Monitoring spatial patterns and changes of surface net radiation in urban and suburban areas using satellite remote-sensing data. International Journal of Remote Sensing. 2017;**38**(4): 1043-1061. DOI: 10.1080/01431161.2016.1275875

[13] Chrysoulakis N. Estimation of the all-wave urban surface radiation balance by use of ASTER multispectral imagery and in situ spatial data. Journal of Geophysical Research. 2003;**108**(D18): 4582. DOI: 10.1029/2003JD003396

[14] Guo Y, Cheng J. Feasibility of estimating cloudy-sky surface longwave net radiation using satellite-derived surface shortwave net radiation. Remote Sensing. 2018;**10**(4). DOI: 10.3390/rs10040596

[15] Amuzurrutia-Valenzuela D, Aguirre-Salado C, Sánchez-Díaz G. ¿Hacia dónde crecerá la ciudad de San Luis Potosí (México) después de 2009? EURE (Santiago). 2015;**41**: 113-117. DOI: 10.4067/S0250-71612015000400006

[16] Lárraga-Lara R. La deforestación urbana y su colaboración a la isla de calor en la Avenida Carranza, San Luis Potosí, México. Revista Caribeña de Ciencias Sociales. 2014;**6**:12. DOI: 10.13140/RG.2.2.21275.92963

[17] Instituto Nacional de Estadística y Geografía. Censo de Población y Vivienda 2020 [Internet]. 2020. Available from: https://www.inegi.org.mx/programas/ccpv/2020/default.html [Accessed: May 28, 2022]

[18] Markham B, Storey J, Morfitt R. Landsat-8 sensor characterization and calibration. Remote Sensing. 2015;7(3): 2279-2282. DOI: 10.3390/rs70302279

[19] Chander G, Markham BL, Helder DL. Summary of current radiometric calibration coefficients for Landsat MSS, TM, ETM+, and EO-1 ALI sensors. Remote Sensing of Environment. 2009;**113**(5):893-903. DOI: 10.1016/j.rse.2009.01.007

[20] Allen RG, Masahiro T, Ricardo T. Satellite-based energy balance for mapping evapotranspiration with internalized calibration (METRIC)—Model. Journal of Irrigation and Drainage Engineering. 2007;**133**(4): 380-394. DOI: 0.1061/(ASCE)0733-9437(2007)133:4(380)

[21] Aguilar-Arias H, Mora-Zamora R, Vargas-Bolaños C. Metodología Para la corrección atmosférica de imágenes aster, rapideye, spot 2 y landsat 8 con el módulo flaash del software envi. Atmospheric correction methodology for aster, rapideye, spot 2 and landsat 8 images with envi flaash module software. Revista Geográfica de América Central [Internet]. 2014;**2**(53):39-59. Available from: https://www.redalyc.org/articulo.oa?id=451744544002

[22] Sheik MP. Spatial-temporal variation of land surface temperature of Jubail Industrial City, Saudi Arabia due to seasonal effect by using Thermal Infrared Remote Sensor (TIRS) satellite data. Journal of African Earth Sciences. 2019;**155**:54-63. DOI: 10.1016/j.jafrearsci.2019.03.008

[23] Rahimi A, Khalil Z, Bouasria A, Mjiri IE, Bounif M. Land surface temperature responses to land use land cover dynamics (district of Taroudant, Morocco). Biology and Life Sciences Forum. 2021;**3**(1). DOI: 10.3390/IECAG2021-09726

[24] Barsi JA, Barker JL, Schott JR. An atmospheric correction parameter calculator for a single thermal band Earth-Sensing Instrument. IGARSS 2003 2003 IEEE International Geoscience and Remote Sensing Symposium Proceedings (IEEE Cat No03CH37477). Toulouse, France: Institute of Electrical and Electronics Engineers Inc.; 2003;5:3014-3016. DOI: 10.1109/ IGARSS.2003.1294665

[25] Ovalle AGC, Tristán AC, Amador-Nieto JA, Putri RF, Zahra RA. Analysing the land use/land cover influence on land surface temperature in San Luis Potosí Basin, México using remote sensing techniques. IOP Conference Series: Earth and Environmental Science. 2021; **686**(1):012029

[26] Instituto Nacional de Ecología y Cambio Climático. [Internet]. Programa de Acción ante el Cambio Climático del Estado de San Luis Potosí 2018. San Luis Potosí, México. 2019. Available from: https://slp.gob.mx/segam/Documentos%20compartidos/ESTUDIOS%20PROGRAMAS%20Y%20PROYECTOS/PEACC%202019%20SLP%20V1.pdf

[27] Santillán-Soto N, García-Cueto R, Haro-Rincón Z, Ojeda-Benítez S, Quintero-Núñez M, Velázquez-Limón N. Radiation balance of urban materials and their thermal impact in Semi-Desert region: Mexicali, México Study Case. Atmosphere. 2015;**6**(10):1578-1589

[28] García Cueto R, Santillán Soto N, Haro Rincón Z, Ojeda Benítez S, Bojórquez Morales G. Parameterization of net radiation in an arid city of northwestern Mexico. Atmosfera. 2015;**28**(2):71-82

[29] Torres Molina LE, Morales S, Carrión LF. Urban Heat Island effects in tropical climate. In: Harun Z, editor. Vortex Dynamics Theories and Applications. London, UK, London, UK: IntechOpen; 2020

[30] Wang L, Wu B, Elnashar A, Zhu W, Yan N, Ma Z, et al. Incorporation of net radiation model considering complex terrain in evapotranspiration determination with Sentinel-2 data. Remote Sensing. 2022;**14**(5):1191

Section 3

Monitoring and Assessment

Chapter 6

Genetic Interaction and Inheritance of Physiobiochemical Traits Can Predict Tolerance of Maize to Water-Deficit Stress

Mozhgan Shirinpour, Ehsan Atazadeh, Ahmad Bybordi, Ashkboos Amini and Hassan Monirifar

Abstract

One of the most critical environmental factors for plant growth is water deficiency and it can be anticipated that climate change will exacerbate this problem in the future. Plants have evolved a variety of different mechanisms at morphological, physiological, cellular, and biochemical levels to overcome water stress conditions. Maize is one of the three leading global cereals, which helps to feed the world. Several biometrical techniques, that is, North Carolina Model, generation mean analysis, diallel and line × tester, are available for genetic analysis. An effective breeding strategy for developing water-deficit tolerant varieties considerably depends on knowledge of the inheritance mechanism of the stress tolerance in maize, high broad-sense heritability and additive genetic variance for the characters which are contributing to drought tolerance. Thus, this study aims toward to explore the inheritance of physiobiochemical traits that lead to increase stress tolerance in maize under the water deficit conditions. This also exhibited a differential pattern of gene action for these traits, suggesting that genotypes possess significant differences for physio-biochemical traits that help to resistance of maize against water deficit stress. Our findings open a door to achieve higher yield of maize under drought stress. These insights might be useful to the plant breeders and farmers for developing water-deficit tolerant maize varieties, and morphological and physio-biochemical markers.

Keywords: gene action, heritability, maize (*Zea mays* L.), physiobiochemical traits, water deficit

1. Introduction

The simulation results suggest that climate variability including storms, flooding, and other extreme weather with increases of temperature may ultimately disrupt crop yield [1]. By century's end, climate change with temperature increases could reduce 11–25% global crop productions [2]. Growth in the temperature and global population can result in shortage of water reserves [3]. Moreover, weather forecast uncertainty will result in decrease of precipitation and increase of evapotranspiration relatively [4]. These factors can lead to drought and reductions in growth and crops yield. One of the main economic solution to increase stability in the production of agricultural products is the genetic modification of plants to withstand abiotic stresses [5]. Therefore, improving crops productivity under deficit irrigation is crucial and needed to ensure food security [6, 7].

As staple food, maize is an important cereal crop and main source of food security which plays a major role in the diets of millions of individuals, fodder, and bioenergy production in the world [8]. Drought stress as the most limitations to productivity of crops than any other abiotic stresses, substantially determines the maize production [9]. Maize is extremely sensitive and vulnerable to water-deficit stress at different growth and development stages that cause great yield reductions during grain fill [10]. Annually, 15–20% of maize production decreases due to climate change and drought. According to FAO statistics [11], maize production in 2016 decreased by 31% in comparison to 2015 due to drought stress and because of many problems of the global climate change and the expansion of maize yield under drought stress conditions. The development of drought-tolerant maize varieties are of high priority in plant breeding and crop science [12].

The knowledge inheritance of various morphological, biochemical, molecular, and physiological mechanisms of the drought maize tolerance is a key breeding strategy and helps the resistance of maize varieties against water-deficit stress [13, 14]. Water-stress tolerance is a complex quantitative trait that is controlled by many microeffective genes [15]. Relevant physiological traits include chlorophyll concentration, relative water content, leaf chlorophyll index, photosynthetic pigments, as well as biochemical traits such as protein content, synthesis of osmolytes, various enzymatic antioxidants, polyphenol oxidase which alters in response to drought and helps resistance of crops against stress [16, 17]. Breeders estimate the effects of genes controlling inheritance of traits in breeding populations, using different mating designs. According to the types of genetic material, the power of estimating additive, dominance, and epistasis gene effects are different [18]. Estimation of variance components of traits (additive, dominance, and epistasis) is important to determine which breeding method can optimize gene action more efficiently to recognize the need to produce hybrids or pure-line varieties [19, 20]. Additionally, levels of the additive effect and dominance degree are very important in designing a plant breeding for improving the trait of interest. Efficiency of selection majority depends on additive genetic variance, the environment, and the genotype × environment interaction effects and this encourages the breeders to understand to what extent the variation is heritable and how much of this variation is usable genetically. Knowledge of interact and gene act determine which breeding system can improve gene action and illustrate the role of this system in the crop plants evolution.

The main objective of this chapter was to determine the genetic control of various physiobiochemical traits which alter in response to water deficit and help the resistance of maize against stress. The knowledge of the physiobiochemical traits can be used to explore new genotypes of maize to increase grain yield under water stress conditions. The genetic component studies, inheritance pattern, and involvement of nonadditive, additive, and maternal genetic effects about various physiobiochemical traits were observed under water-deficit stress conditions. The obtained results will provide a source of potential genetic resources and inheritance patterns, which may be further studied to develop water-deficit tolerance maize varieties and, morphological and physiobiochemical markers.

2. Methods

Various biometrical techniques, that is, North Carolina Model, generation mean analysis, diallel, and line × tester design could be used for understanding of gene action controlling of different plant traits. These methods give information on the importance of average additive and dominance gene effects in determining genotypic values of the generations. Among these, generation mean analysis is the one which determines the type of epistasis (nonallelic gene actions) at digenic level using scaling test, accurately and efficiently [20]. In other words, generation mean analysis model in addition to estimating the genetic parameters viz. mean, additive gene effects, and dominance gene effects, determines three types of nonallelic gene interactions viz. additive × additive, additive × dominance, and dominance × dominance [19]. In this method, the overall average for each trait is shown as follows:

$$Y = m + \alpha[d] + \beta[h] + \alpha 2[i] + \alpha\beta 2[j] + \beta 2[l] \quad (1)$$

where Y: the generation means, m: F∞ metric, d: additive effects, h: dominance effects, i: additive × additive interaction, j: additive × dominance interaction, 1: dominance × dominance interaction, and α, 2αβ and β2: coefficients of genetic parameters. All genetic parameters are tested using a *t*-test for significance. Then additive variance (VA), dominance variance (VD), and environmental variance (VE) are obtained as follows [19]:

$$VA = 2VF2 - VBC1 - VBC2 \quad (2)$$

$$VD = 4(VBC1 + VBC2 - VF2 - VE) \quad (3)$$

$$VE = 0.25(VP1 + VP2 + 2VF1) \quad (4)$$

Broad sense (h^2_{bs}) and narrow sense (h^2_{ns}) heritability are estimated using the following equations:

$$h^2_{bs} = (VA + VD)/(VA + VD + VE); h^2_{ns} = (VA)/(VA + VD + VE) \quad (5)$$

The North Carolina (NC) mating designs permit determination and/or estimation of variance components (additive and dominance components) by using the information from half-sib families. The experimental material of North Carolina designs I, II, and III is developed from F_2 generation as a base material. The design III (NCIII) involves backcrossing the F_2 plants to the two parental inbred lines from which the F_2 were derived. The NCIII design was extended to include a third tester. This third tester is the F_1 from the two parental inbred lines: in this extended form, this design is known as the triple test cross [21]. Line x Tester mating design uses inbred lines as the base population. The design is useful in deciding the relative ability of a number of female and male inbreds to produce desirable hybrid combinations [22]. When the same parents are used as females and males in breeding, the mating design is called diallel. Parents used the range from inbred lines to broad genetic base varieties to clones. The design is the most commonly used in crop plants to estimate general combining ability (GCA) and specific combining ability (SCA) and variances. Analysis of the diallel for GCA and SCA are based on the Model I, which is proposed by Griffing [23]. The GCA/SCA ratio reveals that different characteristics show an additive or nonadditive gene action. AGCA/SCA ratio with a value greater than one indicates additive gene action, whereas a GCA/SCA ratio with a value lower than one indicates dominant gene action. Furthermore, high additive gene action indicates higher heritability and fewer environmental influences [24].

3. Water deficit as the most serious abiotic stress

Among all other abiotic stresses (such as floods, salinity, temperature extremes, heavy metals), water deficit or drought is a significant restricting factor for global agricultural production. Additionally, drought stress is a primary limitation effecting crop yield due to complexity of fresh water limiting and climate change [25, 26]. Water stress occurs when turgor and water potential are reduced to the point where they disorder normal metabolic functions and reproductive capacity of plants [27]. Drought severity depends on several variables, for example, precipitation rate and distribution, evaporative demands and water-maintaining ability of soils [28]. Drought is predicted to become severe owing to global warming, low precipitation, and high evaporation particularly in arid and semiarid regions due to climate change [29]. Moreover, depletion of available water resources, growing world population rise, increasing food demand and climate changes cause water availability less predictable in many areas that exacerbate impacts of drought on global agriculture [30]. Therefore, water scarcity has become a great concern and brought more and more investigations on the drought-resistant crops and knowledge of the water-saving mechanisms of plants under water limitation conditions, especially in arid and semiarid environments [31]. Variety of physiological, biochemical, morphological, and molecular traits of the plants are impaired in water-limiting environments [32].

Water deficit significantly reduces development, grain yield, and yield components of maize at vegetative and reproductive growth stages [33]. Since the heritability of grain yield is low and strongly influenced by the environment, direct selection for it in the different generations is unreliable. Taking this matter into consideration, it is essential for breeders to identify traits that have high heritability, high correlation with grain yield, and low-cost measurement. Then, in the breeding programs, indirect selection for grain yield using these traits could be easier than selection based on direct grain yield [34, 35].

4. Genetic variation of the water-deficit tolerance-associated traits

Understanding about the genetic variation for traits relating to water stress tolerance has great importance in developing breeding program strategies. Depending on the stress duration and severity and the stage of plant development and plant species, water stress causes many morphological, physiological, biochemical, and molecular changes in plants [36]. Plants respond to survive under water deficit condition through the induction of both regulatory and functional sets of genes [37]. The selection of drought-tolerant genotypes and associated traits is globally recognized as an effective strategy to maintain the growth and survival of agricultural crops exposed to future drought periods. A better knowledge of the physiobiochemical traits can be used to create new varieties of crops to increase productivity in water scarcity conditions [38]. Due to the complex genetic basis of water deficit tolerance and poor heritability of the crop yield trait, individual trait components, is more frequently identified and characterized due to its better heritability in replicated experiments [39]. Crop yield becomes especially vulnerable when water deficit stress occurs during the reproductive phase of plant development. Correlated secondary traits are generally easier to measure and show a higher heritability and thus may represent a more suitable target for improving maize response to water stress [40, 41].

4.1 Physiological traits of water-deficit tolerance

Chlorophyll fluorescence is a fast and noninvasive tool used for probing the activity of photosynthesis that can be successfully used for discriminating tolerance of plants to drought stress without damaging the plants [42]. One of the most-often applied chlorophyll fluorescence parameters is F_v/F_m that gives the information about the amount of the light absorbed by photosystem II via chlorophyll [43]. A decrease in the F_v/F_m parameter under water-stress conditions indicates the possibility of inhibiting or destroying the photochemical reaction centers due to the reduction of the electron acceptance and transfer capacity from the chlorophyll complex to photosystem II [44]. This parameter is reliable as a stress indicator in crops which has been documented for photochemical processes [45]. Kalaji et al. [46] concluded that in several studies, fluorescence parameters have been considered as selection tools in plant breeding. They also emphasized the importance of measuring traits related to fluorescence that have a high correlation with yield and especially stress resistance. The previous studies with the genetic analysis of wheat [47] and maize [48, 49] indicated low additive gene effect and high broad sense heritability for F_V/F_m (quantum efficiency of PS II). Contrary, with genome analysis in soybean, moderate magnitude of broad sense heritability has been documented for F_V/F_m [50].

Leaf chlorophyll content as an important physiological parameter plays a major role in evaluating photosynthetic efficiency, crop stresses, and nutritional status for breeding programs [51]. Chlorophyll is an essential pigment for photosynthesis and growth of plants where it converts light energy to chemical energy. In the stress conditions, chlorophyll content of leaves decreases in plants; therefore, it is a useful indicator of plant health [52]. The previous research provided that total chlorophyll content in plants such as wheat, rice, and cotton was controlled by dominance gene effects [53–55]. It is revealed by Said [56] and Shirinpour et al. [57] who reported that generation mean analysis in wheat and maize showed dominance effects for chlorophyll content and chlorophyll index under drought stress, respectively. Whereas in the other research, genetic analysis of chlorophyll index in rapeseed was observed to be under the control of additive gene effects as the main genetic effects for this trait [58]. Meanwhile, Akbari et al. [59] observed the high amount of narrow-sense heritability in the expression of total chlorophyll content and indicated the additive effects that play an important role in the control of this trait. Therefore, selection method could be an effective method to improve the mentioned trait in breeding programs.

The relative water content (RWC) of leaves is a key indicator for the degree of plant cell and tissue wilting and has been suggested as a selection criterion of cultivars for drought stress tolerance [60, 61]. The photosynthetic activity of resistant cultivars to water-deficit stress is higher than sensitive cultivars in high RWC and low-osmotic potential [62]. The reason for this is the ability of most plants to be tolerant to drought stress to reduce water loss through the epidermis of the leaves after closing the stomata or minimizing the open stomata [63]. In addition, the positive and significant correlation of RWC with photosynthesis and grain yield and its high heritability under water stress has made measuring this trait an important indicator in determining the water status of the plant and identifying cultivars resistant to low stress [64, 65]. The relative water content of leaves is one of the traits that lead to increase the stability of yield under drought stress [66]. Under water-deficit stress, a decrease in RWC occurs due to the lack of cell turgor pressure in the leaf and leads to stomatal closure and reduced photosynthetic rate [67]. In fact, the reduction of RWC is the usual response of plants to stress [68]. In some studies, it has been reported that the high percentage of RWC in plants that are exposed to water stress is associated with stress tolerance in plants, and it is considered as an index to determine the difference between cultivars in terms of water-deficit stress tolerance [69, 70]. Moradi et al. [71] by evaluating six lines of maize under normal and drought stress conditions reported that RWC is controlled by the over-dominance effects of genes under both the conditions. In another research by the generation mean analysis of maize, it has been stated that both gene effects of dominance and epistasis play an important role in the genetic control of RWC under control and water-deficit-stress conditions [72]. By studying the gene effect of RWC in sunflower, the main role of additive gene effects in controlling the heritability of this trait has been expressed under water-deficit stress [73, 74]. Naroui Rad et al. [75] observed the importance of both additive and nonadditive gene effects in controlling the inheritance of this trait.

4.2 Biochemical traits of water-deficit tolerance

Plants needed to evolve different mechanisms including osmotic adjustment and antioxidant defense systems, which enhance their capacity to adjust and adapt to water-stress conditions. Water-deficit stress leads to the overproduction of reactive oxygen species (ROS), such as hydrogen peroxide (H_2O_2) and superoxide anion radicals (O_2) which result in plant growth and productivity inhibition [76]. Various enzymatic antioxidants, such as superoxide dismutase, peroxidase, catalase, and ascorbate peroxidase, can be activated to balance between ROS generation and scavenging [77]. Furthermore, accumulation of soluble sugars and proline and changes in protein expression have been reported in maize to increase plant resistance under water-deficit conditions [78]. High narrow-sense heritability (0.53–0.71) has been observed for protein, proline, and soluble sugar concentrations under severe stress conditions. These results indicated that the selection in the parents' inbred lines or early segregating generations could be useful to improve these characters in the maize [79]. Gorji et al. [80] reported that there is moderate and low value of narrow-sense heritability for antioxidant enzymes activities under drought stress and normal condition, respectively. According to Shirinpour et al. [79], the values of narrow-sense heritability for antioxidant enzymes activities of catalase, peroxidases, and polyphenol oxidase were between low to moderate and expressed the major role of dominance variance in the inheritance of these enzymes. For improvement of the antioxidant enzymes, selection in the later generation and utilization of heterosis will facilitate the breeding program.

In corn, proline accumulation due to drought has been reported in the different growth stages for the maintenance of cell turgor and protection of cell structures for improvement under limited water [81]. Under stress conditions, genotypes with high proline content are mainly considered to be tolerant to a number of abiotic stresses suggesting the use of this trait as an index for indirect selection [82, 83]. According to Rahimi [84], proline content displayed over-dominance gene effects by using Haymen's graphical approach in lines of maize. The narrow-sense heritability was 0.14, and this causes the use of heterozygosis and the production of hybrid varieties can be used to breeding of this trait. The same result was reported by Naroui Rad et al. [75] in bread wheat and Khalil et al. [83] in sunflower who found that nonadditive gene effects were predominant for the proline concentration. This result disagrees with Pourmohammad et al. [74] who inferred the presence of additive gene effects for this trait in sunflower. They suggested that selection of proline concentration should be made in the early generations.

Sugars are known as carbon suppliers; therefore, they are considered as an energy source not only for plants but also for most other organisms. Sugar plays a decisive role in the negative osmotic potential in the cytoplasm and, as a result, control the osmotic regulation. Also, they act as osmotic protectors of membranes and proteins and detoxify oxygen-free radicals [85, 86]. They also maintain and stabilize cell membranes under stress conditions [87]. The change in the amount of soluble sugars depends on the type of plant species, and the duration and severity of the drought stress [88]. An increase in the amount of soluble sugars in corn plants has been observed with an increase in the duration and intensity of stress during the low-water period [79, 89]. Accumulation of high amount of sugars such as trehalose, mannitol, sorbitol [90], sucrose, hexose, and raffinose helps the stability of dehydrated membranes and tissues and increases plant tolerance against water stress [91, 92]. In the QTL study for the content of seed-soluble sugars in sweet corn, the importance of both additive and dominance main effects in controlling the inheritance of this trait [93] and the effective role of dominance variance in controlling soluble sugars by examining the genetic analysis of chickpea seeds [94] has been reported.

One of the major biochemical changes that occur due to the decrease in soil moisture in agricultural plants is the change in the amount of production of plant proteins in order to degrade or prevent the synthesis of some of them, as well as the production of stress-specific proteins [95, 96]. The synthesis of stress proteins is the response of the plant to water-deficit-stress conditions. These proteins are soluble in water and therefore, by hydrating the plant's cellular structures, they help to withstand the stress. The synthesis of various transcription factors and stress proteins plays a significant role in the tolerance of plants to stress [97, 98]. Water-deficit stress causes oxidation of carbohydrates, proteins, lipids, and even DNA [99, 100]. In general, water stress reduces the number of cell polysomes in tissues that have less water content, depending on different plant species, and ultimately leads to a reduction in the production of proteins. In addition, the reduction of photosynthesis under drought stress causes a reduction or even a stop of protein production [101, 102]. Enzymes are an important group of intracellular proteins whose activity is inhibited under the influence of water-deficit stress. While some enzymes are stable and their activity may even increase under drought stress [103], Shirinpour et al. [79] using generations mean analysis of hybrids maize SC704 revealed that the high narrow sense heritability and the presence of additive variance for protein content under normal irrigation and water-deficit stress conditions. The authors concluded that the additive gene effects have a major role in controlling of this trait. Then, for improving of the protein content in both normal and water stress conditions, selection in the early segregating generations will be effective. Similar results were observed in the study of Akram et al. [104] and reported the genetic control of protein content by additive and partial dominance effects using bread wheat genetic analysis. These researchers also stated the effectiveness of selection in the early generations for this trait. Meanwhile, Abid et al. [100] indicated the nonadditive gene effects and low narrow sense heritability in the genetic control of protein content in cotton under water-deficit stress.

5. Conclusion

Under recent climatic changes, both the biotic and abiotic stresses are a serious threat for global food security and plant production sustainability. Among the abiotic stresses, water-deficit stress is gaining attention due to its adverse effect on plant growth and development and significant reduction in plant yield causing global food insecurity. Various morphological, biochemical, and physiological mechanisms are affected by water-deficit stress that hampers plant productivity. To tackle the adverse effect of the water stress on plants, certain mechanisms are adopted by the plants which enhance drought tolerance. Genetic improvement of maize for various traits which are contributing to drought tolerance relies primarily on understanding the genetic control of the traits. Knowledge of genetics or genetic control of these traits enables breeders to formulate techniques and strategies for trait selection to bring out the desired genetic improvement in the target traits. Therefore, understanding the genetics of important traits is of paramount importance for rapid and efficient improvement in any crop. This chapter deals with the genetic control of physiobiochemical traits so that effective breeding strategies and selection can be employed to achieve greater progress in maize genetic improvement programs. It can be stated that the genotypes which are maintaining high various physiobiochemical traits such as chlorophyll fluorescence, leaf chlorophyll content, relative water content, various enzymatic antioxidants activities, proline, and sugars accumulation under water-deficit stress are considered tolerant genotypes under deficit moisture stress and used for the selection process. Therefore, these findings might be useful to the maize breeders and farmers for future phenotypic and genotypic association studies. Also, it may be further studied to develop morphological and physiobiochemical markers and will encourage to develop water-deficit tolerance maize varieties. Genetic components depict the involvement of nonadditive and additive genetic effects in the inheritance of various physiobiochemical traits, suggesting that genotypes possess significant differences for these traits. Clearly, additional attention and research on molecular findings of response and resistance of maize generations are needed to identify most resistant crop varieties in field conditions particularly under water-deficit stress.

Author details

Mozhgan Shirinpour[1]*, Ehsan Atazadeh[2], Ahmad Bybordi[3], Ashkboos Amini[4] and Hassan Monirifar[1]

1 Horticulture and Crops Research Department, East Azarbaijan Agricultural and Natural Resources Research and Education Center, AREEO, Tabriz, Iran

2 Department of Plant, Cell, and Molecular Biology, University of Tabriz, Tabriz, Iran

3 Soil and Water Research Department, East Azarbaijan Agricultural and Natural Resources Research and Education Center, AREEO, Tabriz, Iran

4 Seed and Plant Improvement Institute, Agricultural Research, Education and Extension Organization, AREEO, Karaj, Iran

*Address all correspondence to: mozhganshirinpour@gmail.com

References

[1] Khan A, Pan X, Najeeb U, Tan DKY, Fahad S, Zahoor R, et al. Coping with drought: Stress and adaptive mechanisms, and management through cultural and molecular alternatives in cotton as vital constituents for plant stress resilience and fitness. Biological Research. 2018;**51**(1):47

[2] Wing IS, De Cian E, Mistry MN. Global vulnerability of crop yields to climate change. Journal of Environmental Economics and Management. 2021;**109**:102462

[3] Chartres C. Is water scarcity a constraint to feeding Asia's growing population? International Journal of Water Resources Development. 2014;**30**(1):28-36

[4] Abtew W, Melesse A. Climate Change and Evapotranspiration. In: Evaporation and Evapotranspiration. Dordrecht: Springer; 2013. pp. 197-202

[5] Nevo E, Chen G. Drought and salt tolerances in wild relatives for wheat and barley improvement. Plant, Cell and Environment. 2010;**33**(4):670-685

[6] Toumi J, Er-Raki S, Ezzahar J, Khabba S, Jarlan L, Chehbouni A. Performance assessment of AquaCrop model for estimating evapotranspiration, soil water content and grain yield of winter wheat in Tensift Al Haouz (Morocco): Application to irrigation management. Agricultural Water Management. 2016;**163**:219-235

[7] Ket P, Garré S, Oeurng C, Hok L, Degré A. Simulation of crop growth and water-saving irrigation scenarios for lettuce: A monsoon-climate case study in kampong Chhnang, Cambodia. Water. 2018;**10**(5):666

[8] Grote U, Fasse A, Nguyen TT, Erenstein O. Food security and the dynamics of wheat and maize value chains in Africa and Asia. Frontiers in Sustainable Food Systems. 2021;**4**:617009

[9] Aslam M, Maqbool MA, Cengiz R. Drought Stress in Maize (*Zea mays* L.): Effects, Resistance Mechanisms, Global Achievements and Biological Strategies for Improvement. Cham: Springer; 2015

[10] Wang B, Liu C, Zhang D, He C, Zhang J, Li Z. Effects of maize organ-specific drought stress response on yields from transcriptome analysis. BMC Plant Biology. 2019;**19**(1):335

[11] FAOSTAT. Food and Agricultural Organization Statistical Database. Rome, Italy: FAO; 2016. Available from: http://faostatfao.org

[12] Boomsma CR, Vyn TJ. Maize drought tolerance: Potential improvements through arbuscular mycorrhizal symbiosis? Field Crops Research. 2008;**108**(1):14-31

[13] Mittal S, Banduni P, Mallikarjuna MG, Rao AR, Jain PA, Dash PK, et al. Structural, functional, and evolutionary characterization of major drought transcription factors families in maize. Frontiers in Chemistry. 2018;**6**:177

[14] Oosten MJV, Costa A, Punzo P, Landi S, Ruggiero A, Batelli G, et al. Genetics of drought stress tolerance in crop plants. In: Drought Stress Tolerance in Plants. Vol. 2. 2016. pp. 39-70

[15] Blum A. Drought resistance—Is it really a complex trait? Functional Plant Biology. 2011;**38**(10):753-757

[16] Hasibuzzaman ASM, Akter F, Bagum SA, Hossain N, Akter T, Uddin MS. Morpho-physiological mechanisms of maize for drought tolerance. In: Plant Stress Physiology. London, UK: IntechOpen; 2021

[17] Fang Y, Xiong L. General mechanisms of drought response and their application in drought resistance improvement in plants. Cellular and Molecular Life Sciences. 2015;**72**(4):673-689

[18] Muthoni J, Shimelis H. Mating designs commonly used in plant breeding: A review. Australian Journal of Crop Science. 2020;**14**(12):1855-1869

[19] Mather K, Jinks JL. Biometrical Genetics: The Study of Continuous Variation. Chapman and Hall, London : Springer; 1982

[20] Kearsey MJ, Pooni HS. The Genetical Analysis of Quantitative Traits. 2nd ed. London: Chapman and Hall; 2004

[21] Le Clerg E. Significance of Experimental Design in Plant Breeding. Ames: Iowa State University Press; 1966. pp. 243-313

[22] Comstock R, Robinson H. Estimation of average dominance of genes. Heterosis. 1952;**2**:494-516

[23] Griffing B. Concept of general and specific combining ability in relation to diallel crossing systems. Australian Journal of Biological Sciences. 1956;**9**(4):463-493

[24] Hallauer AR, Carena MJ, Miranda Filho J. Quantitative Genetics in Maize Breeding. 3rd ed. New York: Springer-Verlag; 2010. pp. 1-22

[25] Farooq M, Wahid A, Kobayashi N, Fujita D, Basra S. Plant drought stress: Effects, mechanisms and management. Agronomy for Sustainable Development. 2009;**29**(1):185-212

[26] Zhang H, Zhu J, Gong Z, Zhu J-K. Abiotic stress responses in plants. Nature Reviews. Genetics. 2022;**23**(2):104-119

[27] Bijalwan P, Sharma M, Kaushik P. Review of the Effects of Drought Stress on Plants: A Systematic Approach. Switzerland: Preprints.org; 2022

[28] Diatta AA, Fike JH, Battaglia ML, Galbraith JM, Baig MB. Effects of biochar on soil fertility and crop productivity in arid regions: A review. Arabian Journal of Geosciences. 2020;**13**(14):1-17

[29] Badr A, El-Shazly HH, Tarawneh RA, Börner A. Screening for drought tolerance in maize (*Zea mays* L.) germplasm using germination and seedling traits under simulated drought conditions. Plants. 2020;**9**(5):565

[30] Gray SB, Brady SM. Plant developmental responses to climate change. Developmental Biology. 2016;**419**(1):64-77

[31] Sobhanian H, Pahlavan S, Meyfour A. How does proteomics target plant environmental stresses in a semi-arid area? Molecular Biology Reports. 2020;**47**(4):3181-3194

[32] Seleiman MF, Al-Suhaibani N, Ali N, Akmal M, Alotaibi M, Refay Y, et al. Drought stress impacts on plants and different approaches to alleviate its adverse effects. Plants. 2021;**10**(2):259

[33] Cheng M, Wang H, Fan J, Zhang F, Wang X. Effects of soil water deficit at different growth stages on maize growth, yield, and water use efficiency under alternate partial root-zone irrigation. Water. 2021;**13**(2):148

[34] Mwadzingeni L, Shimelis H, Tsilo TJ. Variance components and heritability

of yield and yield components of wheat under drought-stressed and non-stressed conditions. Australian Journal of Crop Science. 2017;**11**(11):1425

[35] Gebregergs G, Mekbib F. Estimation of genetic variability, heritability, and genetic advance in advanced lines for grain yield and yield components of sorghum [*Sorghum bicolor* (L.) Moench] at Humera, Western Tigray, Ethiopia. Cogent Food & Agriculture. 2020;**6**(1):1764181

[36] Chaves MM, Pereira JS, Maroco J, Rodrigues ML, Ricardo CPP, Osório ML, et al. How plants cope with water stress in the field? Photosynthesis and growth. Annals of Botany. 2002;**89**(7):907-916

[37] Bartels D, Sunkar R. Drought and salt tolerance in plants. Critical Reviews in Plant Sciences. 2005;**24**(1):23-58

[38] Yadav S, Sharma KD. Molecular and morphophysiological analysis of drought stress in plants. Plant Growth. Croatia: InTech; 2016. pp. 150-173

[39] Xue Y, Warburton ML, Sawkins M, Zhang X, Setter T, Xu Y, et al. Genome-wide association analysis for nine agronomic traits in maize under well-watered and water-stressed conditions. Theoretical and Applied Genetics. 2013;**126**(10):2587-2596

[40] Liu S, Qin F. Genetic dissection of maize drought tolerance for trait improvement. Molecular Breeding. 2021;**41**(2):1-13

[41] Thirunavukkarasu N, Hossain F, Arora K, Sharma R, Shiriga K, Mittal S, et al. Functional mechanisms of drought tolerance in subtropical maize (*Zea mays* L.) identified using genome-wide association mapping. BMC Genomics. 2014;**15**(1):1-12

[42] Ajigboye OO, Ray RV, Murchie EH. chlorophyll fluorescence on the fast timescale. In: Photosynthesis. Springer; 2018. pp. 95-104

[43] Oukarroum A, El Madidi S, Schansker G, Strasser RJ. Probing the responses of barley cultivars (*Hordeum vulgare* L.) by chlorophyll a fluorescence OLKJIP under drought stress and re-watering. Environmental. Experimental Botany. 2007;**60**(3):438-446

[44] Ahmed IM, Dai H, Zheng W, Cao F, Zhang G, Sun D, et al. Genotypic differences in physiological characteristics in the tolerance to drought and salinity combined stress between Tibetan wild and cultivated barley. Plant Physiology and Biochemistry. 2013;**63**:49-60

[45] Galic V, Franic M, Jambrovic A, Ledencan T, Brkic A, Zdunic Z, et al. Genetic correlations between photosynthetic and yield performance in maize are different under two heat scenarios during flowering. Frontiers in Plant Science. 2019;**10**:566

[46] Kalaji HM, Schansker G, Brestic M, Bussotti F, Calatayud A, Ferroni L, et al. Frequently asked questions about chlorophyll fluorescence, the sequel. Photosynthesis Research. 2017;**132**(1):13-66

[47] Vijayalakshmi K, Fritz AK, Paulsen GM, Bai G, Pandravada S, Gill BS. Modeling and mapping QTL for senescence-related traits in winter wheat under high temperature. Molecular Breeding. 2010;**26**(2):163-175

[48] Šimić D, Lepeduš H, Jurković V, Antunović J, Cesar V. Quantitative genetic analysis of chlorophyll a fluorescence parameters in maize in the field

environments. Journal of Integrative Plant Biology. 2014;**56**(7):695-708

[49] Shirinpour M, Asghari A, Aharizad S, Rasoulzadeh A, Khavari KS. Generation mean analysis of some physiological traits in the hybrid maize cv. SC704 under different water regimes. Journal of Plant Physiology and Breeding. 2020;**10**(1):141-148

[50] Hao D, Chao M, Yin Z, Yu D. Genome-wide association analysis detecting significant single nucleotide polymorphisms for chlorophyll and chlorophyll fluorescence parameters in soybean (*Glycine max*) landraces. Euphytica. 2012;**186**(3):919-931

[51] Fernández-Marín B, Artetxe U, Barrutia O, Esteban R, Hernández A, García-Plazaola JI. Opening Pandora's box: Cause and impact of errors on plant pigment studies. Frontiers in Plant Science. 2015;**6**:148

[52] Zhang H, Ge Y, Xie X, Atefi A, Wijewardane NK, Thapa S. High throughput analysis of leaf chlorophyll content in sorghum using RGB, hyperspectral, and fluorescence imaging and sensor fusion. Plant Methods. 2022;**18**(1):1-17

[53] Naroui Rad MR, Kadir MA, Yusop MR. Genetic behaviour for plant capacity to produce chlorophyll in wheat (*Triticum aestivum*) under drought stress. Australian Journal of Crop Science. 2012;**6**(3):415-420

[54] Song M, Fan S, Pang C, Wei H, Yu S. Genetic analysis of the antioxidant enzymes, methane dicarboxylic aldehyde (MDA) and chlorophyll content in leaves of the short season cotton (*Gossypium hirsutum* L.). Euphytica. 2014;**198**(1):153-162

[55] Makwana R, Patel V, Pandya M, Chaudhari B. Inferences on Magnitude and Nature of Gene Effects for Morpho-Physiological Traits in Rice [*Oryza Sativa* L.]. International Journal of Pure & Applied Bioscience. 2018;**6**(2):1488-1493

[56] Said AA. Generation mean analysis in wheat (*Triticum aestivum* L.) under drought stress conditions. Annals of Agricultural Science. 2014;**59**(2):177-184

[57] Shirinpour M, Asghari A, Atazadeh E, Aharizad S, Rasoulzadeh A. Genetic analysis of grain yield and physiological traits of hybrid maize cv. SC704 under full and water deficit irrigation conditions. Cereal Research Communications. 2021;**49**(2):199-206

[58] Dai W, Girdthai T, Huang Z, Ketudat-Cairns M, Tang R, Wang S. Genetic analysis for anthocyanin and chlorophyll contents in rapeseed. Ciencia Rural. 2016;**46**(5):790-795

[59] Akbari L, Bahraminejad S, Cheghamirza K. Genetic analysis of physiological traits in bread wheat under normal and terminal water-deficit stress conditions. Environmental Stresses in Crop Sciences. 2020;**13**(4):1031-1044

[60] Dhanda S, Sethi G. Tolerance to drought stress among selected Indian wheat cultivars. The Journal of Agricultural Science. 2002;**139**(3):319-326

[61] Chen J, Xu W, Velten J, Xin Z, Stout JJJos, conservation w. Characterization of maize inbred lines for drought and heat tolerance. Journal of Soil and Water Conservation. 2012;**67**(5):354-364

[62] Kaymakanova M, Stoeva N, Mincheva T. Drought stress and its effect on physiological response of sunflower. Central Europe Agriculture. 2008;**9**:749-756

[63] Pirasteh-Anosheh H, Saed-Moucheshi A, Pakniyat H, Pessarakli M. Stomatal responses to drought stress. Water Stress Crop Plants. United States: John Wiley & Sons, Ltd; 2016. pp. 24-40

[64] Schonfeld MA, Johnson RC, Carver BF, Mornhinweg DW. Water relations in winter wheat as drought resistance indicators. Crop Science. 1988;**28**(3):526-531

[65] Merah O. Potential importance of water status traits for durum wheat improvement under Mediterranean conditions. The Journal of Agricultural Science. 2001;**137**(2):139-145

[66] Teulat B, Borries C, This D. New QTLs identified for plant water status, water-soluble carbohydrate and osmotic adjustment in a barley population grown in a growth-chamber under two water regimes. Theoretical and Applied Genetics. 2001;**103**(1):161-170

[67] Kramer PJ, Boyer JS. Water Relations of Plants and Soils. 1st ed. San Diego, California: Academic Press; 1995. p. 495

[68] Pérez-Pérez J, Syvertsen JP, Botía P, García-Sánchez F. Leaf water relations and net gas exchange responses of salinized Carrizo citrange seedlings during drought stress and recovery. Annals of Botany. 2007;**100**(2):335-345

[69] Altinkut A, Kazan K, Ipekci Z, Gozukirmizi N. Tolerance to paraquat is correlated with the traits associated with water stress tolerance in segregating F2 populations of barley and wheat. Euphytica. 2001;**121**(1):81

[70] Keles Y, Öncel I. Growth and solute composition in two wheat species experiencing combined influence of stress conditions. Russian Journal of Plant Physiology. 2004;**51**(2):203-209

[71] Moradi M, Choukan R, Heravan EM, Bihamta MR. Genetic analysis of various morpho-physiological traits in *Zea mays* L. using graphical approach under normal and water stress conditions. Research on. Crops. 2014;**15**(1):62-70

[72] Moharramnejad S, Valizadeh M, Emaratpardaz J. Generation mean analysis in maize (*zea mays* l.) under drought stress. Fresenius Environmental Bulletin. 2018;**27**(4):2518-2522

[73] Rauf S, Sadaqat H, Khan I, Ahmed R. Genetic analysis of leaf hydraulics in sunflower (*Helianthus annuus* L.) under drought stress. Plant, Soil and Environment. 2009;**55**(2):62-69

[74] Pourmohammad A, Toorchi M, Alavikia SS, Shakiba MR. Genetic analysis of yield and physiological traits in sunflower (*Helianthus annuus* L.) under irrigation and drought stress. Notulae Scientia Biologicae. 2014;**6**(2):207-213

[75] Naroui Rad MR, Kadir MA, Yusop MR, Jaafar HZ, Danaee M. Gene action for physiological parameters and use of relative water content (RWC) for selection of tolerant and high yield genotypes in F2 population of wheat. Australian Journal of Crop Science. 2013;**7**(3):407-413

[76] Sun Y, Wang C, Chen HY, Ruan H. Response of plants to water stress: A meta-analysis. Frontiers in Plant Science. 2020;**11**:978

[77] Gill SS, Tuteja N. Reactive oxygen species and antioxidant machinery in abiotic stress tolerance in crop plants. Plant Physiology and Biochemistry. 2010;**48**(12):909-930

[78] Per TS, Khan NA, Reddy PS, Masood A, Hasanuzzaman M, Khan MIR, et al. Approaches in

modulating proline metabolism in plants for salt and drought stress tolerance: Phytohormones, mineral nutrients and transgenics. Plant Physiology and Biochemistry. 2017;**115**:126-140

[79] Shirinpour M, Asghari A, Aharizad S, Rasoulzadeh A, Khavari KS. Genetic interaction and inheritance of biochemical traits can predict tolerance of hybrid maize cv. SC704 to drought. Acta Physiologiae Plantarum. 2020;**42**(7):124

[80] Gorji A, Zonoori Z, Zolnoori M, Jamasbi A. Inheritance of antioxidant activity of triticale under drought stress. Asian Journal of Plant Sciences. 2011;**10**(3):220-226

[81] Sinay H, Arumingtyas EL, Harijati N, Indriyani S. Proline content and yield components of local corn cultivars from Kisar Island, Maluku, Indonesia. International Journal of Plant Biology. 2016;**6**(1):6071

[82] Abbas S, Ahmad S, Sabir S, Shah A. Detection of drought tolerant sugarcane genotypes (Saccharum officinarum) using lipid peroxidation, antioxidant activity, glycine-betaine and proline contents. Journal of Soil Science and Plant Nutrition. 2014;**14**(1):233-243

[83] Khalil F, Rauf S, Monneveux P, Anwar S, Iqbal Z. Genetic analysis of proline concentration under osmotic stress in sunflower (*Helianthus annuus* L.). Breeding Science. 2016;**66**:463-470

[84] Rahimi M. Genetic analysis of Biochemical and Physiological Traits using Haymen's Graphical Approach in Lines and F2 Progenies of Maize (*Zea mays* L.). Plant Genetic Researches. 2021;7(2):1-12

[85] Livingston DP, Hincha DK, Heyer AG. Fructan and its relationship to abiotic stress tolerance in plants. Cellular and Molecular Life Sciences. 2009;**66**(13):2007-2023

[86] Van den Ende W, Valluru R. Sucrose, sucrosyl oligosaccharides, and oxidative stress: Scavenging and salvaging? Journal of Experimental Botany. 2009;**60**(1):9-18

[87] Challabathula D, Bartels D. Desiccation tolerance in resurrection plants: New insights from transcriptome, proteome and metabolome analysis. Frontiers in Plant Science. 2013;**4**:482

[88] Poonam RB, Handa N, Kaur H, Rattan A, Bali S, Gautam V, et al. Sugar signalling in plants: a novel mechanism for drought stress management. In: Ahmad P, editor. Water Stress Crop and Plants: A Sustainable Approach. 1st ed. New York: John Wiley and Sons; 2016. pp. 287-302

[89] Mohammadkhani N, Heidari R. Drought-induced accumulation of soluble sugars and proline in two maize varieties. World Applied Sciences Journal. 2008;**3**(3):448-453

[90] Sharma A, Shahzad B, Kumar V, Kohli SK, Sidhu GPS, Bali AS, et al. Phytohormones regulate accumulation of osmolytes under abiotic stress. Biomolecules. 2019;**9**(7):285

[91] Parida AK, Dagaonkar VS, Phalak MS, Umalkar G, Aurangabadkar LP. Alterations in photosynthetic pigments, protein and osmotic components in cotton genotypes subjected to short-term drought stress followed by recovery. Plant Biotechnology Reports. 2007;**1**(1):37-48

[92] Rosa M, Prado C, Podazza G, Interdonato R, González JA, Hilal M, et al. Soluble sugars: Metabolism, sensing and abiotic stress: A complex network

in the life of plants. Plant Signaling & Behavior. 2009;**4**(5):388-393

[93] Qi X, Zhao Y, Jiang L, Cui Y, Wang Y, Liu B. QTL analysis of kernel soluble sugar content in supersweet corn. African Journal of Biotechnology. 2009;**8**(24):6913-6917

[94] Nassourou MA, Njintang YN, Noubissié TJ-B, Nguimbou RM, Bell JM. Genetics of seed flavonoid content and antioxidant activity in cowpea (*Vigna unguiculata* L. Walp.). The Crop Journal. 2017;**4**(5):391-397

[95] Bewley JD, Larsen KM. Differences in the responses to water stress of growing and non-growing regions of maize mesocotyls: Protein synthesis on total, free and membrane-bound polyribosome fractions. Journal of Experimental Botany. 1982;**33**(3):406-415

[96] Heikkila J, Papp JT, Schultz G, Bewley JD. Induction of heat shock protein messenger RNA in maize mesocotyls by water stress, abscisic acid, and wounding. Plant Physiology. 1984;**76**(1):270-274

[97] Chen L, Wang X. Features and functions of plant moisture-induced protein. Biology Teaching. 2003;**23**(3):503-508

[98] Wahid A, Gelani S, Ashraf M, Foolad MR. Heat tolerance in plants: An overview. Environmental and Experimental Botany. 2007;**61**(3):199-223

[99] Apel K, Hirt H. Reactive oxygen species: Metabolism, oxidative stress, and signal transduction. Annual Review of Plant Biology. 2004;**55**:373-399

[100] Abid MA, Malik W, Yasmeen A, Qayyum A, Zhang R, Liang C, et al. Mode of inheritance for biochemical traits in genetically engineered cotton under water stress. AoB Plants. 2016;**8**:plw008. Epub 2016/02/04

[101] El-Beltagi HS, Mohamed HI, Sofy MR. Role of ascorbic acid, glutathione and proline applied as singly or in sequence combination in improving chickpea plant through physiological change and antioxidant defense under different levels of irrigation intervals. Molecules. 2020;**25**(7):1702

[102] Tayyab N, Naz R, Yasmin H, Nosheen A, Keyani R, Sajjad M, et al. Combined seed and foliar pre-treatments with exogenous methyl jasmonate and salicylic acid mitigate drought-induced stress in maize. PLoS One. 2020;**15**(5):e0232269

[103] Taiz L, Zeiger E. Stress physiology. In: Taiz L, Zeiger E, editors. Plant Physiology. 5th ed. Sunderland: Sinauer Associates, Inc.; 2010. pp. 671-681

[104] Akram Z, Ajmal SU, Kiani AA, Jamil M. Genetic analysis of protein, lysine, gluten and flour yield in bread wheat (*Triticum aestivum* L.). Pakistan Journal of Biological Sciences. 2007;**10**(12):1990-1995

Chapter 7

Recent Trends in the Yield-Nutrient-Water Nexus in Morocco

Terence Epule Epule, Vincent Poirier, Simon Lafontaine, Martin Jemo, Driss Dhiba, Ayoub Kechchour, Soumia Achli, Lahcen Ousayd, Wiam Salih and Perez Lionnel Kemeni Kambiet

Abstract

Climate change is impacting environmental systems including agriculture. In Morocco, declining precipitation and increasing temperatures are negatively impacting crop yields. Consequently, crop yields in Morocco are now dependent on nutrient and water management. Most studies have focused on experimentation through fertilizer application and irrigation without any attention to the intrinsic linear relationships that exist between crop yields, fertilizers, and agricultural water withdrawal. The time series agricultural water withdrawal data were collected from AQUASTAT for the period 1990-2022 while data on nitrogen, phosphorous, and potash fertilizers were collected from FAOSTAT. Yield data for maize, barley, sorghum, and wheat were also collected from FAOSTAT. The data were analyzed using two machine learning models fitted through multiple linear regression. The key results show that for the three fertilizers, phosphates tend to have the strongest impacts and cause changes in crop yield as seen in the context of wheat. When both fertilizers and agricultural water withdrawal are fitted against yield, agricultural water withdrawals tend to have a strong relationship with yields. This work has helped us to identify which crops and management options need to be valorized in terms of increased access to nutrients and water.

Keywords: agriculture, water withdrawal, water management nutrient management, trends, multiple linear regression

1. Introduction

Across Africa, agriculture is essentially rainfed; however, due to climate change and variability, precipitation is not often sufficient for crop production. Consequently, alternative sources of water need to be harnessed for crop production. The recent Intergovernmental Panel for Climate Change Report (IPCC) [1] has noted that across

Africa, temperatures will continue to increase while the pattern of precipitation will continue to be variable regionally. For example, the various Representative Concentration Pathway (RCP) scenarios show that in terms of precipitation, north Africa including Morocco and southern Africa will continue to witness declines in precipitation, especially during the growing season of crops while west, central and east Africa will witness high intensity and poorly distributed precipitation. To ensure food security for over nine billion people by 2050, agricultural production at a global scale must increase by between 70 and 100% [2]. If care is not taken, global production as per recent projections for the period 2006–2050 will decline from 2.2 to 1.1% [3]. In Africa, during the same period, the annual growth rate of food production is projected to decline from 3 to 2.1% [3]. These downward trends in crop production are likely to have devastating effects; some of these constraints are crop yield related [4–7].

Amidst these constraints imposed by climate on agriculture, across Africa and around the world, crop production has become increasingly based on water withdrawals/irrigation and fertilisation [8]. Currently, only about 15% of Moroccan agriculture is irrigated while about 85% is dependent on rainfall. Within this context of insufficient rainfall and an arid environment, crop yields are thus subjected to enormous stress [8]. Food insecurity and lagging food production in Africa are shifting attention to irrigation. Irrigation and fertilisation are among key investments and technical inputs that are needed to revamp crop production in Africa [8]. Therefore, there is an increased emphasis on the valorisation of water withdrawal and fertilisation in a bid to adapt to the limits imposed by climate change. Water withdrawal represents the total amount of water extracted from river, soil moisture, ground water and precipitation and used to enhance crop productivity. To respond to these stressors, communities, governments and other organisations across Africa are making efforts to make African agriculture resilient through sustainable withdrawal of water and fertilisation [8, 9]. Across Africa, organisations, such as the African Development Bank (AfDB), United Nations Reduction of Emissions from Deforestation and Forest Degradation (UN-REDD+) and Office Chérifien des Phosphates (OCP) Africa/Foundation, are now having programmes that aim at enhancing agriculture by the valorisation of water withdrawal and fertilisation [10–12].

Morocco is located on the North-West edge of the African continent, between latitudes 21°N and 36°N and longitudes 1°W and 17°W. The country has a total area of nearly 711,000 km^2. This includes 2934 km of coast on the Atlantic Ocean to the West and 512 km of coast on the Mediterranean Sea to the North. It borders Algeria to the East and South-East and Mauritania to the South-West. According to the latest 2014 census, its population is estimated at nearly 34 million people. Morocco is characterised by a wide variety of topographies ranging from mountains and plateaus to plains, oases and Saharan dunes. For this reason, the country experiences diverse climatic conditions with large spatial intra- and inter-annual variability of precipitation. Morocco faces irregular rain patterns, cold spells and heat waves increasingly resulting in droughts, which significantly affects agriculture [13, 14].

Morocco's policy to modernise its agriculture and make it profitable for small- and medium-sized farmers and for the Moroccan economy in general is outlined in the "Green Morocco Plan" (GMP) that was established in 2008–2018 [13, 14], now replaced by the "Generation Green Plan" (GGP) to cover the next decade [15]. These plans have established goals to assist farmers to access water for agriculture as well as other agricultural inputs such as fertilisers (Agence Pour le Développement Agricole (i.e. the Agricultural Development Agency (ADA)) [16, 17]. Invariably, these policies have had positive effects on the Moroccan agricultural economy envisioned through increased agricultural production epitomised through greater access to fertilisers, water for irrigation and increased mastery of the irrigation process.

Due to this emphasis on water withdrawal and fertilisation, research interest has also been tilted towards this direction. However, most of the studies have focused on aspects of agricultural management in the context of experimental agriculture. The goal has so far been to investigate through process-based models of how various scenarios of irrigation and fertilisation impact crop response [18–24]. To the best of our knowledge, this is the first study that uses historical data to assess recent trends in the yield of maize, barley, sorghum and wheat at a national scale in Morocco. The national scale approach adopted is essentially aspatial and seeks to provide insights into how national historical yields of the concerned crops respond to nutrient and water management. Therefore, this work assesses recent trends in the relationship between yield (dependent variable) and fertilisers (nitrogen, phosphate and potash-independent variables) on the one hand and between crop yield (dependent variable) and water withdrawals and fertilisers (independent variable) on the other. This approach is used to determine the effect that fertilisers and agricultural water withdrawal have on the selected crops to determine where emphasis can be made to improve food production. Therefore, this work seeks to assess at a national scale the recent trends in the relationships between the yields of the various crops and how they interact with aspects of nutrient and water management. This approach will improve understanding of these recent trends at the national scale. It will also go a long way in providing vital information necessary for closing yield gaps.

2. Materials and methods

2.1 Data collection

To be able to assess the recent trends in the crop yield, fertiliser and water withdrawal nexus, historical data on the key variables (crop yield, agricultural water withdrawal, nitrogen, phosphate and potash) were collected. The methodological steps employed in this study are presented in **Figure 1**. The collected time series data were selected for the period 1990–2020 for which data were available. For the years 2012–2022, the missing data were obtained by linear interpolation (see Section 2.3 below). Less than an insignificant 10% of the data were missing. Yield data were collected on maize, barley, sorghum and wheat (hectograms per hectare per year (hg/ha/year)) and were collected from FAOSTAT (FAO's agricultural database) [25]. The historical data on fertilisers for nitrogen, phosphate and potash (tons/year) were also collected from FAOSTAT (FAO's fertiliser statistics) [25]. The fertiliser data presented here represent fertiliser data used in agriculture in Morocco and reported by the government of Morocco to the FAO). Agricultural water withdrawal data were downloaded from AQUASTAT (FAO's irrigation and water withdrawal database) [26]. The data from FAOSTAT and AQUASTAT are observed data that are collected from individual countries and reported by the FAO. In cases of missing data, the FAO estimates the missing data but, in most cases, the data are observed records. Agricultural water withdrawal is used here as a proxy for irrigation. It includes surface water, ground water and non-conventional water that has been used for agriculture. The agricultural water withdrawal data are often reported by governments to the FAO.

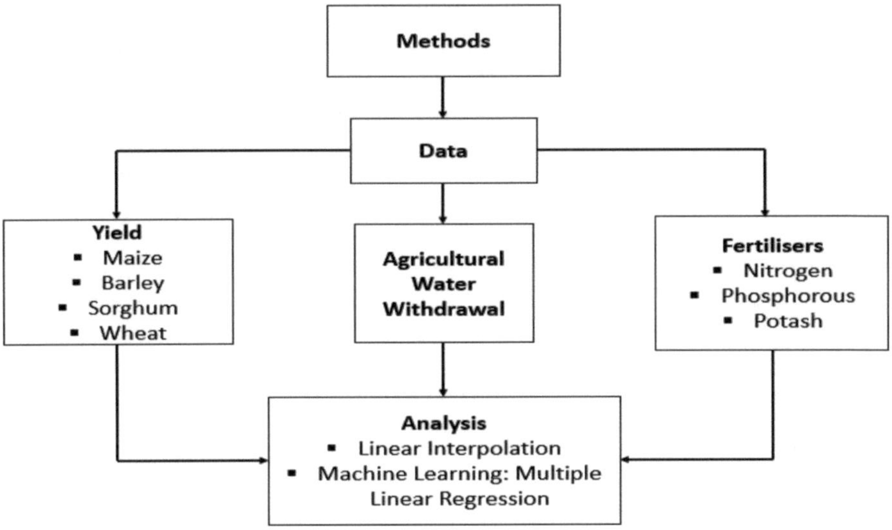

Figure 1.
Methodological steps employed.

2.2 Data analysis

The time series historical data were collected for the period 1990–2020. However, except for agricultural water withdrawal that had complete data from 1961 to 2020, all the other variables (fertilisers and yields) covered the data from 1990 to 2020. Therefore, the analysis was performed with the data that spanned the period 1990–2020. The data were analysed using the Statistical Package for the Social Sciences (SPSS) software while the graphs were produced using excel software. The inferential statistics used included simple linear regression, multiple linear regression, coefficient of determination (R^2), p-values (significant at ($p \leq 0.05$), t-values, linear interpolation and the related standard errors.

2.3 Linear interpolation

For homogeneity, the linear interpolation approach was used to estimate the missing data for yield and fertilisers for the period 2021–2022. The linear interpolation procedure used is defined below (Eq. (1)). The linear interpolation method is often used to estimate the unknown values or missing data through the known values or available data.

$$\text{Yyrnr} = \text{y1} + ((\text{x–x1})/(\text{x2} - \text{x1}))^{*}\,(\text{y2} - \text{y1}) \tag{1}$$

where Yyrnr represents the unknown values or missing data for maize, barley, sorghum and wheat yield as well as missing values for nitrogen, phosphate and potash fertilisers, x represents the known values for maize, barley, sorghum and wheat yields as well as missing values for nitrogen, phosphate and potash fertilisers, x1 and y1 are the coordinates that are below the known x value and x2 and y2 are the coordinates that are above the known x value.

Once linear interpolation was finalised, a complete dataset for maize, barley, sorghum and wheat yields as well as missing values for nitrogen, phosphate and potash fertilisers were obtained covering the period 1990–2022.

2.4 Multiple linear regression

To assess the relationship between yield, agricultural water withdrawal and fertilisers, multiple linear regression was used to identify which independent variable impacts yield the most. Two regressions were fitted for each crop (Eqs. (2)–(9)), one based on the relationship between yield and nitrogen, phosphate and potash and one based on the relationship between yield, fertilisers and agricultural water withdrawal. This procedure was repeated for each crop as follows:

2.4.1 Maize

$$\mathrm{Ymaize\_yf} = \mathrm{a} + \mathrm{bX1} + \mathrm{bX2} + \mathrm{bX3} + \in \tag{2}$$

where X1, X2, X3 (for nitrogen, phosphate and potash, respectively) are the explanatory variables and Ymaize_yf is the dependent variable (maize yield). The slope of the line is b, and a is the intercept (the value of y when x = 0) and ∈ is the error term.

$$\mathrm{Ymaize\_ywf} = \mathrm{a} + \mathrm{bX1} + \mathrm{bX2} + \in \tag{3}$$

where X1, X2 (agricultural water withdrawal and fertilisers, respectively) are the explanatory variables and Ymaize_ywf is the dependent variable. The slope of the line is b, and a is the intercept (the value of y when x = 0) and ∈ is the error term.

2.4.2 Barley

$$\mathrm{Ybarley\_yf} = \mathrm{a} + \mathrm{bX1} + \mathrm{bX2} + \mathrm{bX3} + \in \tag{4}$$

where X1, X2, X3 (nitrogen, phosphate and potash, respectively) are the explanatory variables and Ybarley_yf is the dependent variable (barley yield). The slope of the line is b, and a is the intercept (the value of y when x = 0) and ∈ is the error term.

$$\mathrm{Ybarley\_ywf} = \mathrm{a} + \mathrm{bX1} + \mathrm{bX1} + \in \tag{5}$$

where X1, X2 (agricultural water withdrawal and fertilisers, respectively) are the explanatory variables and Ybarley_ywf is the dependent variable (barley yield). The slope of the line is b, and a is the intercept (the value of y when x = 0) and ∈ is the error term.

2.4.3 Sorghum

$$\mathrm{Ysorghum\_yf} = \mathrm{a} + \mathrm{bX1} + \mathrm{bX2} + \mathrm{bX3} + \in \tag{6}$$

where X1, X2, X3 (nitrogen, phosphate and potash, respectively) are the explanatory variables and Ysorghum_yf is the dependent variable (sorghum yield). The slope of the line is b, and a is the intercept (the value of y when x = 0) and ∈ is the error term.

$$Ysorghum\_ywf = a + bX1 + bX2 + \in \quad (7)$$

where X1, X2 (agricultural water withdrawal and fertilisers, respectively) are the explanatory variables and Ysorghum_ywf is the dependent variable (sorghum yield). The slope of the line is b, and a is the intercept (the value of y when x = 0) and ∈ is the error term.

2.4.4 Wheat

$$Ywheat\_yf = a + bX1 + bX2 + bX3 + \in \quad (8)$$

where X1, X2, X2 (nitrogen, phosphate and potash, respectively) are the explanatory variables and Ywheat_yf is the dependent variable (wheat yield). The slope of the line is b, and a is the intercept (the value of y when x = 0) and ∈ is the error term.

$$Ywheat\_ywf = a + bX1 + bX2 + \in \quad (9)$$

where X1, X2 (agricultural water withdrawal and fertilisers, respectively) are the explanatory variables and Ywheat_ywf is the dependent variable (wheat yield). The slope of the line is b, and a is the intercept (the value of y when x = 0) and ∈ is the error term.

3. Results

3.1 Fertiliser use in Moroccan agriculture

In terms of the use of nitrogen, phosphate and potash fertilisers in African agriculture, Morocco is among the highest users as can be seen in **Figure 2**. For example, in the context of nitrogen fertilisers used in agriculture, Morocco records about 1,155332.34 tons together with other Africa countries like Egypt, Nigeria, Ethiopia and South Africa (**Figure 2a**). In terms of phosphate fertiliser usage in agriculture, Morocco still ranks among the top users in Africa. The country records about 410,513.54 tons together with countries like Algeria, Egypt, Nigeria, Ethiopia, Kenya and South Africa (**Figure 2b**). In the context of potash, Morocco still ranks among the highest users in Africa, using about 355,225.81 tons in its agriculture together with other leading countries such as Egypt, Nigeria and South Africa (**Figure 2c**).

3.2 The evolution of crop yields, fertilisers (nutrients) and agricultural water withdrawals in Morocco

In the context of the four crops under consideration in this study (maize, barley, sorghum and wheat), the evolution over time is one that has been highly variable and fluctuating between 1990 and 2022 (**Figure 3a**). Despite this observation, wheat

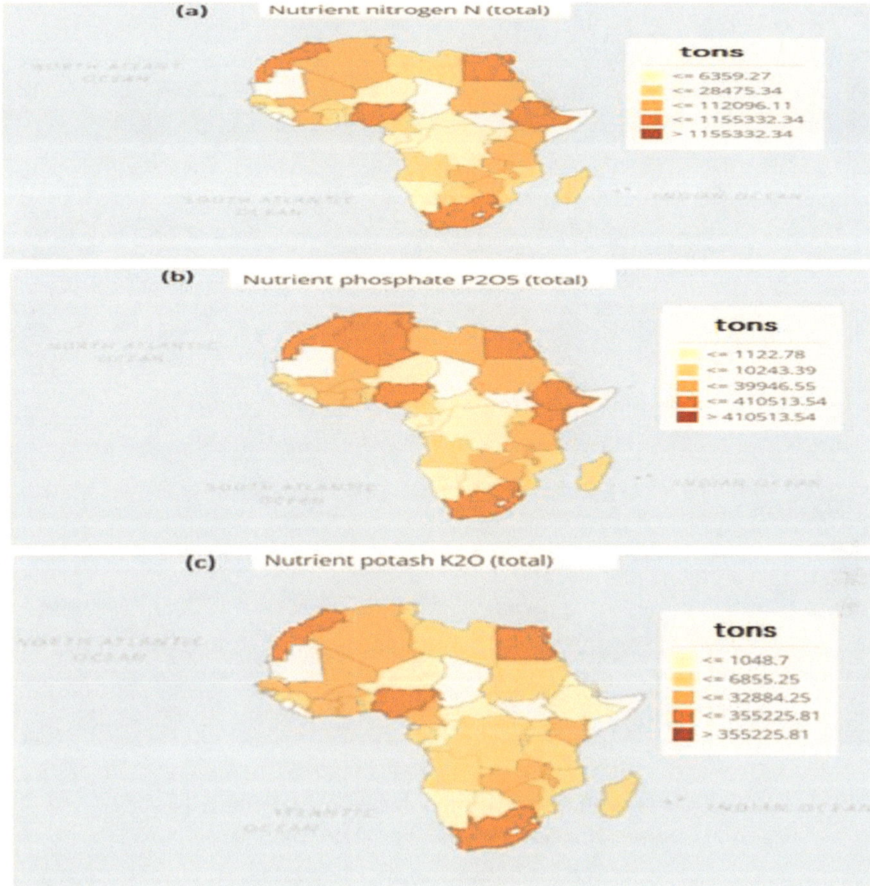

Figure 2.
Agricultural fertilisers (tons) used in Africa (1990–2020). (a) Nitrogen, (b) phosphate and (c) potash. Source: Developed by authors from FAOSTAT.

records the highest yield during this period followed by barley, maize and sorghum (**Figure 3a**). In terms of fertilisers/nutrients used in agriculture in Morocco, the results are consistent with those reported in **Figure 2a, b** and **c**. For example, in Morocco, though also slightly variable as yields, fertilisers/nutrients are generally dominated by nitrogen fertilisers that record the highest values throughout the series; this is followed by phosphate and potash, respectively (**Figure 3b**). In terms of agricultural water withdrawal, the historical data show a slight decline between 1990 and 2022 (**Figure 3c**).

3.3 Scatter plots and linear regression outputs of the relationship between fertilisers/nutrients and maize yields in Morocco

The scatter plots depict the linear relationship between maize yields as the dependent variable and fertilisers as the independent variable. In the context of nitrogen fertilisers used in maize cultivation, it can be observed that an R^2 of 0.0014 (0.14%) is obtained. This implies that only 0.14% of changes in maize yield can be explained by changes in nitrogen fertiliser application (**Figure 4a**). In terms of phosphate fertilisers

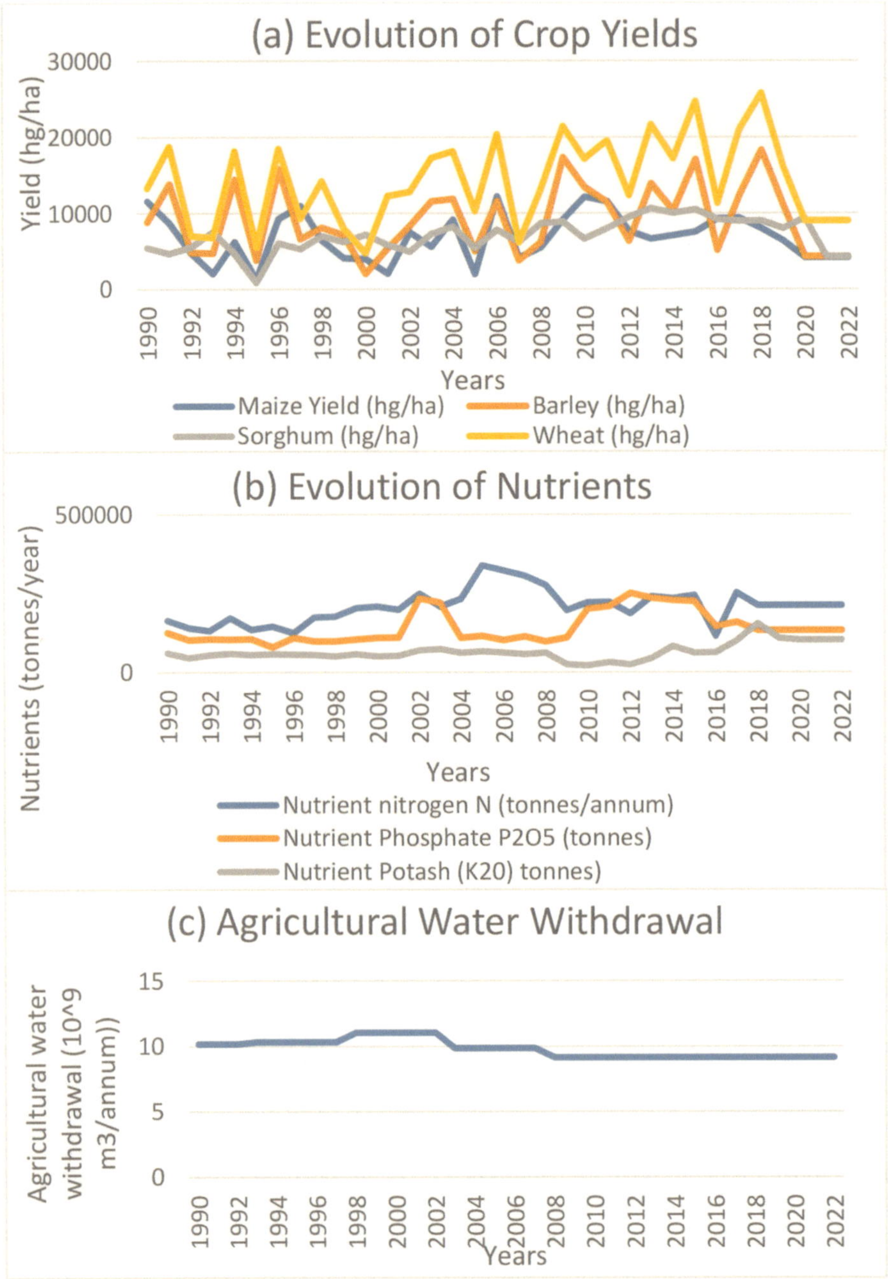

Figure 3.
Evolution of (a) yields, (b) fertilisers/nutrients and (c) water withdrawal for irrigation in Morocco.

used in maize cultivation, it can be observed that an R^2 of 0.07 (7%) is obtained. This implies that only 7% of changes in maize yield can be explained by changes in phosphate fertiliser application (**Figure 4b**). Considering potash fertilisers used in maize cultivation, it can be observed that an R^2 of 0.041 (4.1%) is obtained. This implies that only 4% of changes in maize yield can be explained by changes in potash fertiliser application (**Figure 4c**). From these linear trends, it can be observed that the

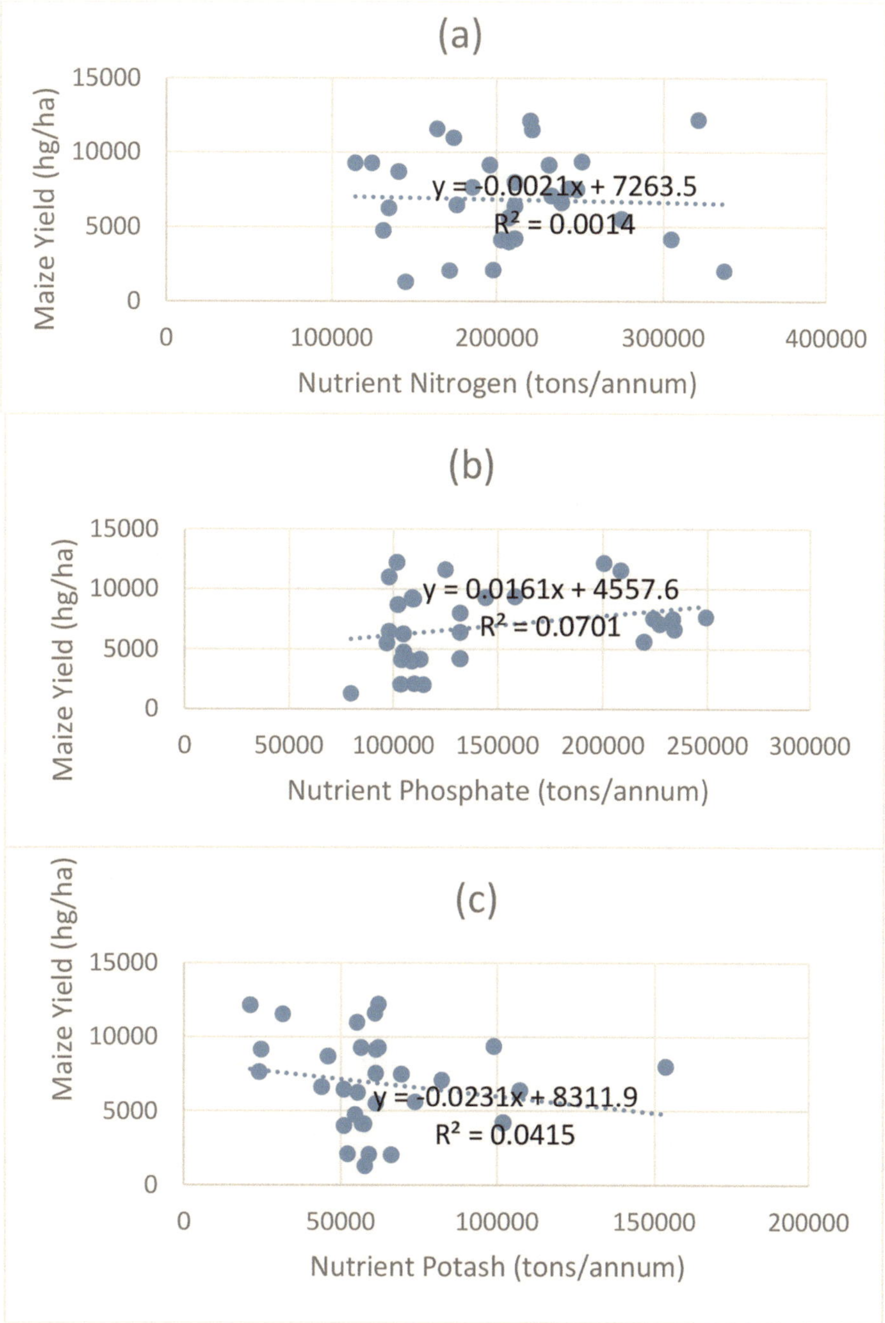

Figure 4.
Scatter plots of maize yield against (a) nutrient nitrogen, (b) nutrient phosphate and (c) nutrient potash.

changes in maize yields cannot be explained by changes in fertiliser application. However, phosphate fertilisers seem to outbid the other nutrients as they record relatively higher R^2 of 7%. Phosphate fertilisers tend to explain more of the changes in maize yield as depicted by the R^2 of 7%, a statistic that is relatively higher than those recorded for nitrogen and potash.

The results from the scatter plots (**Figure 4a, b** and **c**) are consistent with multiple linear regression outputs (**Table 1**). This is observed, as phosphate fertilisers tend to record the lowest p-value of 0.15 and the highest t-value of 1.44. This is followed by potash (p-value = 0.34, t-value = −0.95) and lastly nitrogen (p-value = 0.72, t-value- = −0.35). Even though nitrogen represents higher levels of application in Moroccan agriculture, when it comes to maize, phosphates tend to explain more of the changes in maize yield than the other fertilisers. When the linear relationship between maize yield as the dependent variable and agricultural water withdrawal and fertilisers as the independent variables is considered, it is observed that agricultural water withdrawal records the lower p-value (0.11) and the highest t-value (−1.63) (**Table 2**). In relative terms, agricultural water withdrawal tends to explain more of the changes in maize yield when compared to fertilisers.

3.4 Scatter plots and linear regression outputs of the relationship between fertilisers/nutrients and barley yields in Morocco

The scatter plots depict the linear relationship between barley yield as the dependent variable and fertiliser as the independent variable. In the context of nitrogen fertilisers used in barley cultivation, it can be observed that an R^2 of 0.0005 (0.05%) is obtained. This implies that only 0.05% of changes in barley yield can be explained by changes in nitrogen fertiliser application (**Figure 5a**). In terms of phosphate fertilisers used in barley cultivation, it can be observed that an R^2 of 0.076 (7.6%) is obtained. This implies that only 7.6% of changes in barley yields can be explained by changes in phosphate fertiliser application (**Figure 5b**). Considering potash fertilisers used in barley cultivation, it can be observed that an R^2 of 0.0002 (0.02%) is obtained. This implies that only 0.02% of changes in barley yield can be explained by changes in potash fertiliser application (**Figure 5c**). Phosphate fertilisers outbid the other nutrients as they record the highest R^2 of 7.6% and the lowest p-values of 0.11. Phosphate fertilisers tend to explain more of the changes in barley yield as depicted by the R^2 of 7.6%, a statistic that is much higher than those recorded for nitrogen and potash.

The results from the scatter plots (**Figure 5a, b** and **c**) are consistent with linear regression outputs (**Table 3**). This is observed as phosphate fertilisers tend to record the lowest p-value of 0.11 and the highest t-value of 1.64. This is followed by nitrogen (p-value = 0.63, t-value = −0.48) and lastly potash (p-value = 0.77, t-value = 0.29). Even though nitrogen represents higher levels of application in Moroccan agriculture, when it comes to barley, phosphates tend to explain more of the changes in barley yields. Still, none of these nutrients correlates significantly with barley yields. When the linear relationship between barley yields as the dependent variable and agricultural water withdrawal and fertilisers as the independent variables is considered, it is observed that agricultural water withdrawal records the lower p-value (0.15) and the highest t-value (−1.44) (**Table 4**).

3.5 Scatter plots and linear regression outputs of the relationship between fertilisers/nutrients and sorghum yields in Morocco

In the context of nitrogen fertilisers used in sorghum cultivation, it can be observed that an R^2 of 0.075 (7.5%) is obtained. This implies that only 7.5% of changes in sorghum yield can be explained by changes in nitrogen fertiliser application (**Figure 6a**). In terms of phosphate fertilisers used in sorghum cultivation, it can be observed that an R^2 of 0.23(23%) is obtained. This implies that only 23% of changes in

	Coefficients	Standard error	t Stat	P-value	Lower95%	Upper95%	Lower95.0%	Upper 95.0%
Intercept	6591.479609	2684.042776	2.455803	0.020287	1101.99576	12080.963	1101.995763	12080.9635
X1: Nitrogen	−0.003664534	0.010421889	−0.35162	0.727666	−0.0249797	0.0176506	−0.02497969	0.01765062
X2: Phosphate	0.015876546	0.01095039	1.449861	0.157827	−0.0065195	0.0382726	−0.00651952	0.03827261
X3: Potash	−0.01932795	0.020283127	−0.95291	0.348512	−0.0608116	0.0221557	−0.0608116	0.0221557

Table 1.
Linear regression outputs of the relationship between maize yield and fertilisers.

	Coefficients	Standard error	t Stat	P-value	Lower95%	Upper95%	Lower95.0%	Upper95.0%
Intercept	21890.29571	10197.37344	2.14666	0.040026	1064.4808	42716.111	1064.48081	42716.11061
X1: Water withdrawal	−1410.92082	864.2597603	−1.63252	0.113024	−3175.9747	354.13308	−3175.97472	354.1330848
X:2 Fertilisation	−0.00292644	0.007058972	−0.41457	0.681406	−0.0173428	0.0114899	−0.01734278	0.011489907

Table 2.
Linear regression outputs of the relationship between maize yield, water withdrawal and fertilisers.

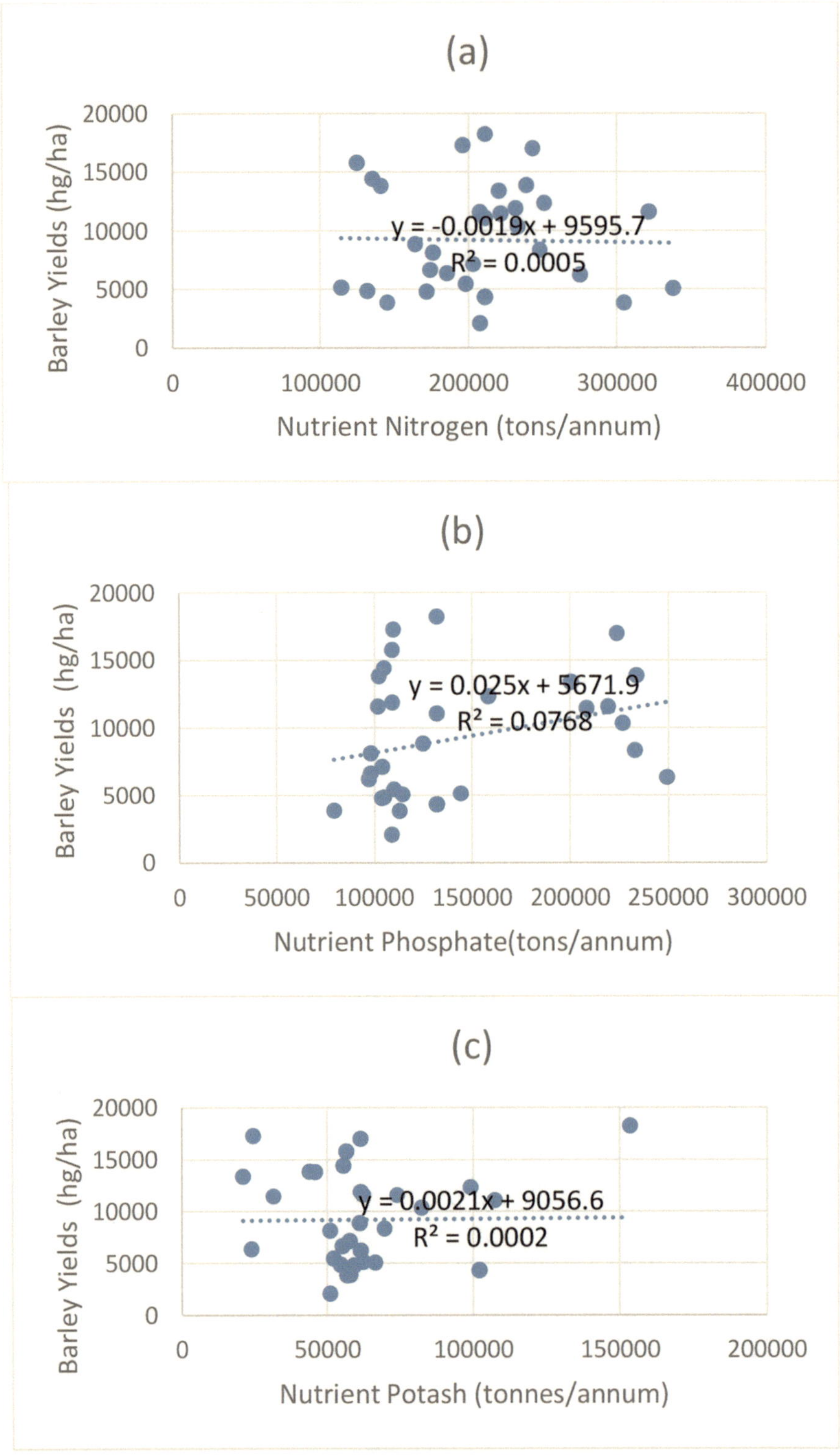

Figure 5.
Scatter plots of barley yield against (a) nutrient nitrogen, (b) nutrient phosphate and (c) nutrient potash.

	Coefficients	Standard error	t Stat	P-value	Lower95%	Upper95%	Lower95.0%	Jpper95.0%
Intercept	6389.667811	4031.559477	1.584912	0.123832	−1855.797	14635.133	−1855.7971	14635.133
X1: Nitrogen	−0.00760109	0.015654171	−0.48556	0.630925	−0.039617	0.0244153	−0.0396175	0.0244153
X2: Phosphate	0.027003673	0.016448005	1.64176	0.111442	−0.006636	0.0606436	−0.0066363	0.0606436
X3: Potash	0.008948938	0.030466218	0.293733	0.771054	−0.053361	0.0712593	−0.0533615	0.0712593

Table 3.
Linear regression outputs of the relationship between barley yield and fertilisers.

	Coefficients	Standard error	t Stat	P-value	Lower 95%	Upper 95%	Lower95.0%	Upper 95.0%
Intercept	26914.105	15137.24273	1.778006	0.085541	−4000.2689	57828.4789	−4000.2689	57828.47889
X1: Water Withdrawal	−1849.11365	1282.929359	−1.44132	0.159851	−4469.2049	770.97764	−4469.2049	770.9776401
X2: Fertilisation	0.00106867	0.01047852	0.101986	0.919446	−0.0203313	0.02246866	−0.0203313	0.022468657

Table 4.
Linear regression outputs of the relationship between barley yield, water withdrawal and fertilisers.

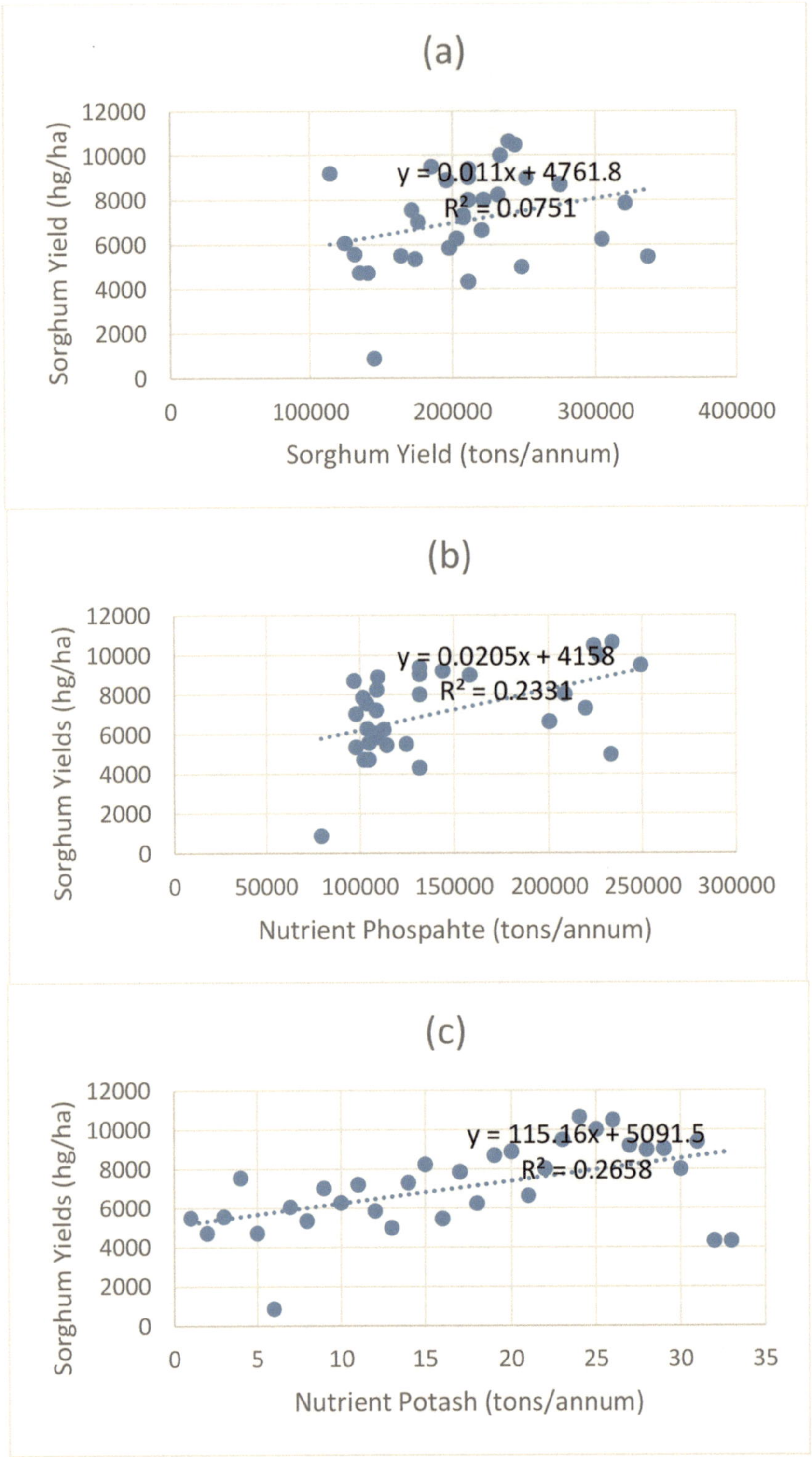

Figure 6.
Scatter plots of sorghum yield against (a) nutrient nitrogen, (b) nutrient phosphate and (c) nutrient potash.

sorghum yields can be explained by changes in phosphate fertiliser application (**Figure 6b**). Considering potash fertilisers used in sorghum cultivation, it can be observed that an R^2 of 0.26 (26%) is obtained. This implies that only 26% of changes in sorghum yield can be explained by changes in potash fertiliser application (**Figure 6c**). Phosphate fertilisers outbid the other nutrients as they record a relatively higher R^2 of 23% and the lowest p-values of 0.009. Phosphate fertilisers tend to explain more of the changes in sorghum yield as depicted by the R^2 of 23%, a statistic that is much higher than those recorded for nitrogen and potash.

The results from the scatter plots (**Figure 6a, b** and **c**) are consistent with linear regression outputs (**Table 5**). This is observed as phosphate fertilisers tend to record the lowest relative p-value of 0.009 and the highest t-value of 2.78. This is followed by nitrogen (p-value = 0.29, t-value = −1.06) and lastly potash (p-value = 0.68, t-value- = −0.40). Even though nitrogen represents higher levels of application in Moroccan agriculture, when it comes to sorghum, phosphates tend to impact ~~maize~~ yields more than the other fertilisers. When the linear relationship between sorghum yields as the dependent variable and agricultural water withdrawal and fertilisers as independent variables is considered, it is observed that agricultural water withdrawal records the lower p-value (0.025) and the highest t-value (−2.35) (**Table 6**). This depicts the fact that agricultural water withdrawal has a more significant relationship with sorghum yield when compared to fertiliser application.

3.6 Scatter plots and linear regression outputs of the relationship between fertilisers/nutrients and wheat yields in Morocco

In the context of nitrogen fertilisers used in wheat cultivation, it can be observed that an R^2 of 0.015 (1.5%) is obtained. This implies that only 1.5% of changes in wheat yield can be explained by changes in nitrogen fertiliser application (**Figure 7a**). In terms of phosphate fertilisers used in wheat cultivation, it can be observed that an R^2 of 0.03 (3%) is obtained. This implies that only 3% of changes in wheat yields can be explained by changes in phosphate fertiliser application (**Figure 7b**). Considering potash fertilisers used in wheat cultivation, it can be observed that an R^2 of 0.0002 (0.02%) is obtained. This implies that only 0.02% of changes in wheat yields can be explained by changes in potash fertiliser application (**Figure 7c**). Also, though most of the relationships are weak, phosphate fertilisers seem to outbid the other nutrients as they record a relatively higher R^2 of 3% and the lowest p-values of 0.05. Phosphate fertilisers tend to explain more of the changes in wheat yield as depicted by the R^2 of 3%, a statistic that is much higher than those recorded for nitrogen and potash.

The results from the scatter plots (**Figure 7a, b** and **c**) are consistent with linear regression outputs (**Table 7**). This is observed as phosphate fertilisers tend to record the lowest p-value of 0.05 and the highest t-value of 2.04. This is followed by nitrogen (p-value = 0.78, t-value = −0.26) and lastly potash (p-value = 0.81, t-value = −0.23). Even though nitrogen represents higher levels of application in Moroccan agriculture, when it comes to wheat, phosphates tend to impact wheat yields more than the other fertilisers. When the linear relationship between wheat yields as the dependent variable and agricultural water withdrawal and fertilisers as independent variables are considered, it is observed that agricultural water withdrawal records the lower p-value (0.17) and the highest t-value (−1.39) (**Table 8**). This depicts the fact that agricultural water withdrawal though generally weak has a relatively more significant relationship with wheat yield when compared to fertiliser application.

	Coefficients	Standard error	t Stat	P-value	Lower 95%	Upper 95%	Lower 95.0%	Upper 95.0%
Intercept	2526.199866	1697.373588	1.488299	0.147467	−945.3189	5997.71864	−945.31891	5997.718642
X1: Nitrogen	0.007045388	0.006590744	1.068982	0.293895	−0.006434	0.02052497	−0.0064342	0.020524973
X2: Phosphate	0.019310401	0.006924965	2.78852	0.009252	0.0051473	0.03347354	0.005147257	0.033473544
X3: Potash	0.005243352	0.012826935	0.408777	0.685707	−0.020991	0.03147738	−0.02099068	0.031477381

Table 5.
Linear regression outputs of the relationship between sorghum yield and fertilisers.

	Coefficients	Standard error	t Stat	P-value	Lower 95%	Upper95%	Lower95.0%	Upper95.0%
Intercept	15953.27167	6053.100472	2.635554	0.013167	3591.191298	28315.35204	3591.1913	28315.352
X1: Water withdrawal	−1207.806837	513.0194746	−2.35431	0.025299	−2255.53238	−160.081294	−2255.5324	−160.081294
X2: Fertilisation	0.007163757	0.004190164	1.70966	0.097655	−0.0013937	0.015721214	−0.0013937	0.01572121

Table 6.
Linear regression outputs of the relationship between sorghum yield, water withdrawal and fertilisers.

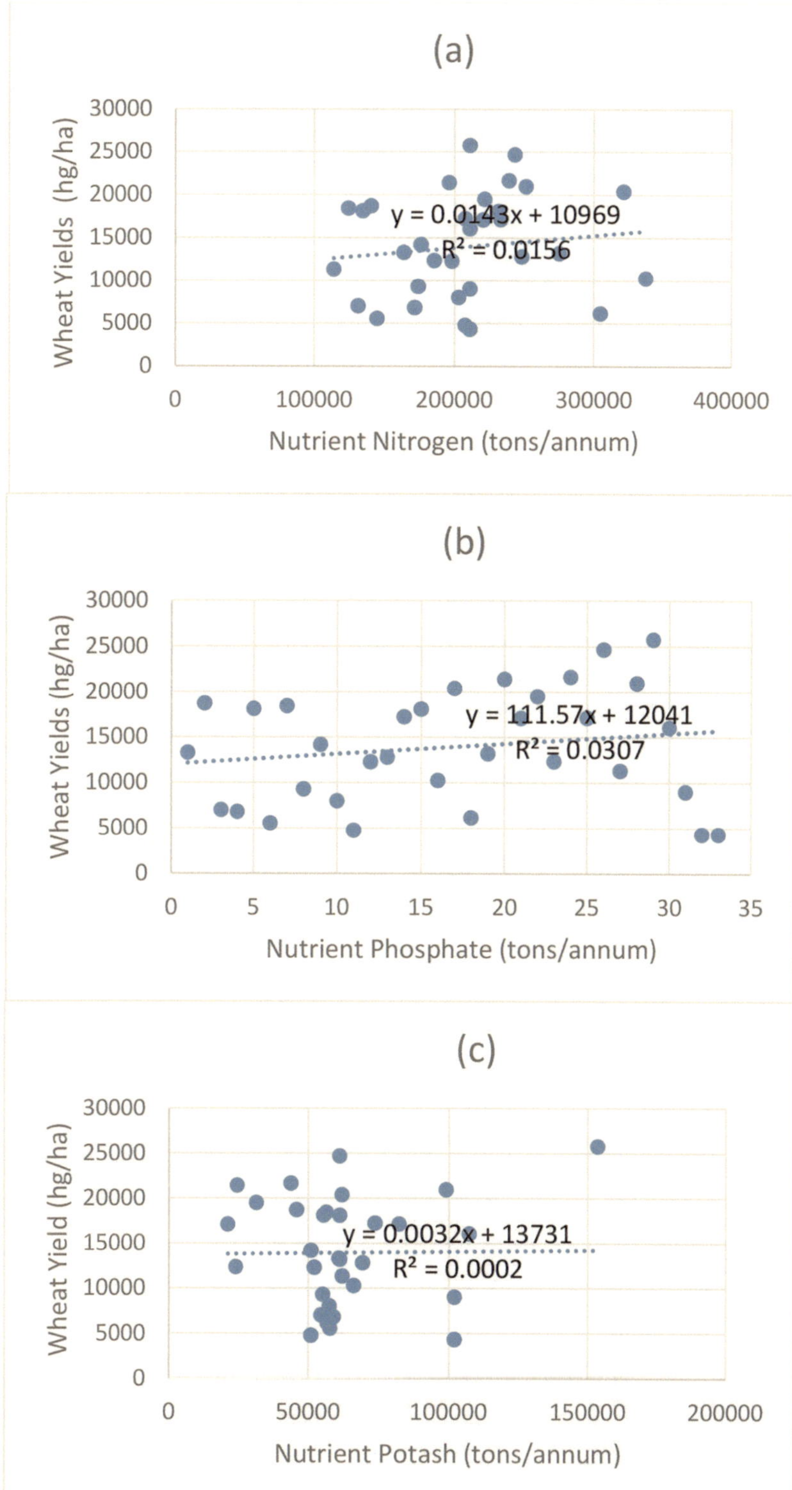

Figure 7.
Scatter plots of wheat yield against (a) nutrient nitrogen, (b) nutrient phosphate and (c) nutrient potash.

	Coefficients	Standard error	t Stat	P-value	Lower 95%	Upper 95%	Lower 95.0%	Upper95.0%
Intercept	6032.704941	5254.512938	1.1481	0.26031	−4713.98067	16779.3906	−4713.9807	16779.391
X1: Nitrogen	0.005490829	0.020402786	0.269122	0.789742	−0.03623755	0.04721921	−0.0362376	0.0472192
X2: Phosphate	0.043772796	0.021437425	2.041887	0.050351	−7.166E-05	0.08761725	−7.166E-05	0.0876173
X3: Potash	0.009229943	0.039707993	0.232445	0.817825	−0.07198202	0.09044191	−0.071982	0.0904419

Table 7.
Linear regression outputs of the relationship between wheat yield and fertilisers.

	Coefficients	Standard error	t Stat	P-value	Lower 95%	Upper 95%	Lower95.0%	Upper95.0%
Intercept	31606.3683	19681.80113	1.605868	0.118779	−8589.232	71801.9686	−8589.232	71801.96865
X1: Water Withdrawal	−2321.3313	1668.095106	−1.39161	0.174275	−5728.036	1085.37339	−5728.036	1085.37339
X2: Fertilisation	0.01242275	0.013624419	0.9118	0.369143	−0.015402	0.04024752	−0.015402	0.040247522

Table 8.
Linear regression outputs of the relationship between wheat yield, water withdrawal and fertilisers.

4. Discussion

This work has found that of the three fertilisers under study and for all the crops included, phosphate fertilisers tend to have a relatively stronger linear relationship with yields. Wheat records the most significant relationship with fertilisers and more specifically in the case of phosphates (p-value = 0.05). Phosphate fertilisers are obtained from phosphorus rocks that are essential in crop growth and food cultivation. Phosphorus fertilisers constitute a strong element in Moroccan agriculture. Phosphorus is essential found in just a handful of countries around the world, with Morocco being the number one producer of phosphate fertilisers followed by China [27, 28]. About 70% of the global distribution of phosphate rocks will be depleted within the next 100 years. Morocco, with >77% of global phosphorus rock reserves, will need to increase production by 700% to be able to meet its deficits by 2075 [27, 28]. Invariably, Morocco's reserves accounted for almost all the increases in the global reserves of phosphate rocks [28, 29]. Morocco will have to increase its production to be able to meet the local and international demand for phosphate fertilisers. This work has assessed the recent trends in the linear relationships between the yields of the concerned crops and fertilisers and water withdrawal. This approach is important because as in the case of Morocco, as an arid country in which agriculture is mainly rainfed, water and nutrient management are key drivers of crop yields. However, there is need for more analysis with several variables on the climatic and non-climatic spectra to be able to better understand these trends. Perhaps therefore, the key contribution here is that yield prediction based on a few variables as in the case of this work does not often yield good outputs as yields are complex and generally impacted by a complex combination of several factors. In the current scientific literature, it remains unclear how these components of nutrient and water management play out in determining crop yields. For example, there are no national scale studies that have integrated components of nutrient and water management in yield prediction. Existing studies have either focused on nutrients or water management mostly in the context of irrigation [30]. A better understanding of this relationship will play a crucial role in determining how well yield gaps can be closed through nutrient and water management. In other words, this work sets the pace for a better understanding of the combinations and the need for more research for improved insights into such relationships on nutrient and water or irrigation that can be used to close future yield gaps. By understanding the linear trends between yield and nutrients or water, it becomes possible to identify where the linear trends are weak and which combinations of irrigation and nutrients can be used to close such gaps. The only other study that comes close to this is the work published in 2012 in Nature by Muller et al. [30] as it examines the use of nutrient and water management strategies to determine yield gaps using essential non-linear regression models at a global scale. Our work builds on the same strategy but focuses essentially on Morocco for which the global study did not dwell in detail.

In Morocco, it is important to however understand that there exists a spatial variation in crop yield as are climatic conditions. Normally, from south to north, crop yield increases for most crops. This has been explained by the south–north spatial variations in climate. This is seen as a straddle from south to north shows increased precipitation and lower temperatures due to the presence of the Atlantic Ocean and the Mediterranean Sea. Furthermore, sociodemographic data have also shown that due to its European and temperate influence, the north has higher literacy and lower poverty rates. This mix of climatic and non-climatic variables plays an important role in determining crop yields. The results have shown that in most cases, the changes in yield are explained by relatively small changes in fertilisation as depicted by the R^2. There is a need to identify how much, when and where nutrients and water are needed to improve crop yields. This is especially true as many of the farmers involved are essentially small-scale farmers who do not have access to nutrients and irrigation. Wheat, on the other hand, has a relatively more significant relationship with phosphate fertilisers when compared to the other crops because the crop has been more valorised over the years as it is not just a major staple crop, but it also plays an important economic role in the country. Therefore, this work helps us to understand which crops need more valorisation in the context of access and use of nutrients and water, insights that can help in the formulation of agricultural policy.

It is now evident from previous scholarship that the relatively stronger impact that phosphate fertilisers have on crop yield in Morocco in general and especially on wheat when compared to the other crops is highly tied to the country's dominance in the production of phosphate fertilisers, which is also tied to its huge phosphate rock reserves [27, 29]. Morocco is often described as the Saudi Arabia of phosphorus because it is blessed with huge phosphate rock reserves. It has been argued that comparing Morocco with Saudi Arabia in terms of Morocco's phosphate rock reserves is very simplistic as Morocco would be better compared with all OPEC (Organisation of the Petroleum Exporting countries) put together [31–33].

The current global reserves of phosphate rocks may last for an additional 300–400 years. By this time, most countries would have exhausted their stocks. This global figure is hugely impacted by Morocco with a rock-phosphate ratio of close to 2000 years. Globally, the countries that currently have rock-phosphate ratios of less than 100 years are responsible for about 70% of the global production and continuous depletion will result in huge deficits [28]. China and the United States account for over 50% of the global production and these might be depleted within the next 60 years at the current rates of extraction. The delicate global phosphorus intricate situation that surrounds phosphate extraction is going to intensify as Morocco has an ever-greater share of global phosphate production [29, 34]. This has implications for food and phosphorus security unless new phosphate reserves can be accessed. This will further mean that the world will be more reliant on Morocco [28, 35].

In terms of the relationship between yield as a dependent variable and agricultural water withdrawal and fertilisers as independent variables, agricultural water withdrawals are observed to have a stronger relationship. Here, it can be concluded that agricultural water withdrawal as a proxy for irrigation in agriculture plays a very important role. As a management option for agriculture, water withdrawal plays a keen role in filling the gaps created by unreliable precipitation. Farmers in most arid and semi-arid countries and regions often depend on precipitation as the main source of water for their crops. However, due to climate change and other anthropogenically induced stressors such as unsustainable systems of farming, there is often not enough precipitation for crop growth for systems that are basically rainfed. The recent IPCC Sixth Assessment Report confirms this assertion when it notes that most of north Africa including Morocco will continue to witness declining precipitation and rising temperature under various Representative Concentration Pathways (RCP) scenarios [36, 37]. The government of Morocco, through the Green Morocco Plan (GMP) and the Generation Green Strategy (GGS), have subsidised irrigation to make it accessible for farmers at various scales. Irrigation is therefore a significant driver of agriculture in Morocco.

Within the pessimistic climate conundrum that necessitates irrigation and fertilisation, such water withdrawals are often based on surface water resources, groundwater resources and non-convectional water resources. The mean precipitation records about 140 billion cubic meters per year [38, 39]. Evapotranspiration triggers a leakage of about 118 billion cubic meters per year. Natural water potential is estimated at about 22 billion cubic meters per year. In Morocco, the amount of economically exploitable water is 80% of the available water resources, revealing the constraints in water resources and the difficulties associated with their utilisation for agriculture [39]. For example, the hydrological regime of all basins is characterised by rife interannual variability marked by wet and dry sequences, interspaced with years of intense drought stress. The Ouergha basin, for example, is among the most productive basins in Morocco by virtue of its surface water flow variations of 2.5 billion cubic meters per year. Thus, the large regional differences in rainfall also trigger large surface water flow variations that vary from a few million cubic meters (MCM) with most arid basins in Morocco such as the Saharan basins, to about one billion cubic meters per year for the most water-rich basins. The Sebou basin in the north of Morocco, for example, holds 30% of surface and groundwater resources in Morocco. Although it represents only 6% of the total area of Morocco, 18% of the country's population lives within its frontiers [40, 41]. Internal water reallocations are applied to reduce the shortages in each river basin. About 0.3 million cubic meters per year is transferred from the Oum er-Rbia basin to dry areas in Tensift, essentially to sustain irrigation. Similarly, another 0.16 million metric cubic meters per year is transferred from the Sebou and Oum er-Rbia basins to support Bouregreg's domestic water needs in other parts of Morocco. Ground water, on its part, accounts for 20% of all water resources in Morocco. Aquifers cover about 80,000 km^2, about 10% of the national territory. Total groundwater withdrawal is estimated at about 3170 million cubic meters per year. However, the huge water withdrawal stress here is exhibited as about 4.2 billion cubic meters per year are extracted (higher than the rate of recharge) [30, 42]. On a final note, non-conventional water withdrawal is governed by the National Water Plan, which aims to address issues of gaps between water demand and water supply through *inter alia* desalinisation of sea water. The establishments of plants to this effect have enabled the desalination of nearly 515 million cubic meters per year [30, 42, 43].

The generally weak relationship between yield and fertilisers on the one hand and water withdrawal on the other hand shows us that crop yield is not that simple. Crop yields are determined by a complex interplay of several variables including precipitation, temperature, soils, fertilisers, irrigation, crop pests and diseases and livestock. Isolating yield and a few independent variables may provide insights into how many of the changes in yield are explained by changes in fertilisers and irrigation, but it is good to caution that several variables often impact crop yields.

Phosphorus can also be released from compost, green manures and animal manures. These elements contain mineralised phosphorus and micronutrients that are easier for plants to use. These alternative ways of adding phosphorus to increase soil fertility are good but cannot independently match the phosphorus that is obtained from phosphorus rocks [30, 40–43]. Also, combining phosphorus rock with green manure crop buckwheat might have significant benefits but phosphate rock deposits will remain the main source of phosphate fertilisation. Incubated phosphorus rocks have been observed to increase phosphorus uptake in buckwheat crop but the buckwheat residues did not however enhance the yield of the next crop. It has however been observed that using green manure crops on different soils might produce more positive results, but the scale does not match natural phosphorus. Green manure crop like legumes might also perform better [30, 40–43].

In the case of inorganic fertilisers, the phosphorus is removed from the rock using acids, rendering the phosphorus more soluble and easier to absorb. The problem now is to employ the same technique in organic agriculture to see how organic farms can enhance the update of phosphorous—one of the three key micronutrients together with potash and nitrogen for plant growth. This is even more challenging as micronutrients have no substitutes and their absence might hamper plant growth [30, 40–43]. In fact, inorganic phosphorus can be obtained through farm-level recycling of organic materials such as composts, green manure and animal manures. Adding green manure residues to soils can increase mineralisation rates of phosphorus but, at the same time, low concentrations of residues often do limit crop demand. Using green manure crop species that have high phosphorus uptake can overcome this limitation. Buckwheat, for example, absorbs concentrations of phosphorus beyond its own needs such that excess of it is left in the soil for future use [30, 40–43]. This has evidently shown us that, the future of phosphorus fertilisation depends on natural phosphorus rocks and Morocco will continue to play an important role in the agricultural food-chain in Morocco.

5. Conclusions

This study has shown that though fertiliser application and water withdrawals have a weak relationship with crop yields (indicating the need for valorisation), these inputs nevertheless potentially constitute key variables that can trigger a veritable agricultural revolution in Moroccan agriculture that needs to be investigated further. More specifically, this work has shown that phosphate fertilisers have strong impact on the yields of all the crops, especially wheat, relative to nitrogen and potash. With the reality of the depletion of phosphorus rocks across the world in the frame of the huge dependence of agricultural systems on phosphate fertilisers, several questions come to mind. For example, how can land use policy be streamlined to ensure that global agricultural systems continue to have the phosphates they need to thrive? What are the potential alternative sources of phosphorus away from the huge phosphorus deposits in Morocco? If the Moroccan phosphorus monopoly is maintained, what are the likely actions needed to ensure access to phosphate fertiliser for the rest of Africa without extreme high prices? These among others are some of the questions that come to mind when the likely future trends in the use and stock of phosphorus resources are considered.

However, this study focuses mainly on fertiliser use and irrigation; in reality, crop production is more complex than simplified here, as it is often impacted by a complex interplay of several climatic and non-climatic variables. Potential areas of further research however would estimate yield gaps using the relationship between actual and attainable yield towards closing yield gaps as well as the relationship closing the latter through nutrient and water management. The use of machine learning-based approaches to determine the drivers of crop yields with several climatic and non-climatic drivers will go a long way in providing a balanced understanding of crop yields.

Author details

Terence Epule Epule[1]*, Vincent Poirier[1], Simon Lafontaine[1], Martin Jemo[2], Driss Dhiba[3], Ayoub Kechchour[3], Soumia Achli[3], Lahcen Ousayd[3], Wiam Salih[3] and Perez Lionnel Kemeni Kambiet[3]

1 Unité de recherche et développement en agroalimentaire (URDAAT), Université du Québec en Abitibi-Témiscamingue (UQAT), Notre-Dame-du-Nord, Québec, Canada

2 Agrobiosciences, Mohammed 6 Polytechnic University, Ben Guerir, Morocco

3 International Water Research Institute, Mohammed 6 Polytechnic University, Ben Guerir, Morocco

*Address all correspondence to: terenceepule.epule@uqat.ca

References

[1] Intergovernmental Panel for Climate Change (IPCC). Impacts, Adaptation and Vulnerability. IPCC Sixth Assessment Report: Summary for Policy Makers. Available from: https://www.ipcc.ch/report/ar6/wg2/ [Accessed: 7 October 2022]

[2] Dubois O. The State of the world's Land and Water Resources for Food and Agriculture: Managing Systems at Risk. Available from: http://www.fao.org/3/i1688e/i1688e00.pdf [Accessed: 7 October 2022]

[3] Bruinsma J. The resource outlook to 2050. In: Expert Meeting on "How to Feed the World in 2050"; 24–26 June 2009; Rome, Italy. Rome: FAO. Available from: http://www.fao.org/3/a-ak971e.pdf [Accessed: 10 October 2022]

[4] Sanchez PA. Soil fertility and hunger in Africa. Science. 2002;**295**:2019-2020

[5] Thomson H, Berrang-Ford L, Ford JD. Climate change and food security in Sub-Saharan Africa: A systematic literature review. Sustainability. 2010;**2**:2719-2733

[6] Sasson A. Food security for Africa: An urgent global challenge. Agriculture & Food Security. 2012;**1**:1-16

[7] Bruinsma J. The resource outlook to 2050. In: Expert Meeting on "How to Feed the World in 2050"; 24–26 June 2009; Rome, Italy. Rome: FAO. Available from: http://www.fao.org/3/a-ak971e.pdf [Accessed: 16 September 2022]

[8] Svendsen M, Ewing M, Msangi S. Measuring Irrigation Performance in Africa. Vol. 894. Washington DC: International Food Policy Research Institute (IFPRI); 2009

[9] Sakaki M, Koga K. An effective approach to sustainable small-scale irrigation developments in Sub-Saharan Africa. Paddy and Water Environment. 2013;**11**(1):1-14

[10] Salami A, Arawomo DF. Empirical analysis of agricultural credit in Africa: Any role for institutional factors. African Development Bank Group. 2013;**192**:1-28

[11] Dooley E, Chapman S. Climate-smart agriculture, and REDD+ implementation in Kenya. Environmental Management. 2014;**4**:2-2

[12] Becker A, Benabderrazik K, Bertsimas D, Chtinna N, El Majdoub N, Mahboubi EM, et al. Toward global food security: Transforming OCP through analytics. INFORMS Journal on Applied Analytics. 2022;**52**(1):90-107

[13] Faysse N. The rationale of the Green Morocco Plan: Missing links between goals and implementation. The Journal of North African Studies. 2015;**20**(4):622-634

[14] Saidi AS, Diouri M. Food self-sufficiency under the Green-Morocco Plan. Journal of Experimental Biology and Agricultural Sciences. 2017;**5**(Spl-1-SAFSAW):33-40

[15] Mathez A, Loftus A. Endless modernisation: Power and knowledge in the Green Morocco Plan. Environment and Planning E: Nature and Space, 25148486221101541. 2022;**6**(1):87-112

[16] Agence Pour le Développement Agricole (ADA). Bilan: Pan Maroc Vert. Available from: https://www.ada.gov.ma/fr/fondements [Accessed: 7 October 2022]

[17] Agence Pour le Développement Agricole (ADA). Nouvelle Stratégie du Secteur Agricole. Available from: https://

www.ada.gov.ma/fr/nouvelle-strategie-du-secteur-agricole [Accessed: 7 October 2022]

[18] Kharrou MH, Le Page M, Chehbouni A, Simonneaux V, Er-Raki S, Jarlan L, et al. Assessment of equity and adequacy of water delivery in irrigation systems using remote sensing-based indicators in semi-arid region, Morocco. Water Resources Management. 2013, 2013;**27**(13):4697-4714

[19] Jarlan L, Abaoui J, Duchemin B, Ouldbba A, Tourre YM, Khabba S, et al. Linkages between common wheat yields and climate in Morocco (1982–2008). International Journal of Biometeorology. 2014;**58**(7):1489-1502

[20] Bregaglio S, Frasso N, Pagani V, Stella T, Francone C, Cappelli G, et al. New multi-model approach gives good estimations of wheat yield under semi-arid climate in Morocco. Agronomy for Sustainable Development. 2015;**35**(1): 157-167

[21] Bishaw Z, Yigezu YA, Niane A, Telleria RJ, Najjar D. Political Economy of the Wheat Sector in Morocco: Seed Systems, Varietal Adoption, and Impacts. Beirut, Lebanon: International Center for Agricultural Research in the Dry Areas; 2019. Available from: http://inrameknes.info/wp-content/uploads/2019/03/PoliticalEconomyWheatSectorMorocco_WEB_ISBN.pdf [Accessed 7 September 2022]

[22] Bouasria A, Ibno Namr K, Rahimi A, Ettachfini EM. Geospatial assessment of soil organic matter variability at Sidi Bennour District in Doukkala Plain in Morocco. Journal of Ecological Engineering. 2021;**22**(11):120-130

[23] Bouras EH, Jarlan L, Er-Raki S, Balaghi R, Amazirh A, Richard B, et al. Cereal yield forecasting with satellite drought-based indices, weather data and regional climate indices using machine learning in Morocco. Remote Sensing. 2021;**13**(16):3101

[24] Bouras E, Jarlan L, Khabba S, Er-Raki S, Dezetter A, Sghir F, et al. Assessing the impact of global climate changes on irrigated wheat yields and water requirements in a semi-arid environment of Morocco. Scientific Reports. 2019;**9**(1):1-14

[25] FAO. AQUASTAT: Core Database. Available from: AQUASTAT <ah ref=""http://www.fao.org/nr/water/aquastat/data/query/index.html?lang=en""target=""_blank"">Website</a> [Accessed: 17 August 2022]

[26] FAO. FAOSTAT: Core Database. Available from: https://www.fao.org/faostat/en/#data/RFN [Accessed: 17 August 2022]

[27] Vaccari DA. Phosphorus: A looming crisis. Scientific American. 2009;**300**: 54-59

[28] Cooper J, Lombardi R, Boardman D, Carliell-Marquet C. The future distribution and production of global phosphate rock reserves. Resources, Conservation and Recycling. 2011;**57**:78-86

[29] Smit AL, Bindraban PS, Schröder JJ, Conijn JG, Van Der Meer HG. Phosphorus in Agriculture: Global Resources Trends and Developments. Wageningen, The Netherlands: Plant Research International B.V; 2009

[30] Mueller ND, Gerber JS, Johnston M, Ray DK, Ramankutty N, Foley JA. Closing yield gaps through nutrient and water management. Nature. 2012; **490**(7419):254-257

[31] Cordell D. The Story of Phosphorus. Sustainability Implications of Global

Phosphorus Scarcity for Food Security. Linkoping University; 2010

[32] Cordell D, Drangert J-O, White S. The story of phosphorus: Global food security and food for thought. Global Environmental Change. Linkoping University. 2009;**19**:292-305

[33] Finley M, Economics US. BP Statistical Review of World Energy June 2017. Annual Report. United Kingdom: Baker Institute; 2017

[34] Van Kauwenbergh SJ. World Phosphate Rock Reserves and Resources. International Fertilizer Development Center (IFDC); International Fertiliser Développement Coopertaion. 2010

[35] Brunner PH. Substance flow analysis as a decision support tool for phosphorus management. Journal of Industrial Ecology. 2010;**14**(6):870-873

[36] Chen S, Chen W, Wu R, Yu B, Song L. Performance of the IPCC AR6 models in simulating the relation of the western North Pacific subtropical high to the spring northern tropical Atlantic SST. International Journal of Climatology. 2021;**41**(4):2189-2208

[37] Zhongming Z, Linong L, Xiaona Y, Wangqiang Z, Wei L. AR6 Climate Change: The Physical Science Basis. IPCC; 2021

[38] Schyns JF, Hoekstra AY. The added value of water footprint assessment for national water policy: A case study for Morocco. PLoS One. 2021;**9**(6):e99705, 2014m

[39] Toumi J, Er-Raki S, Ezzahar J, Khabba S, Jarlan L, Chehbouni A. Performance assessment of AquaCrop model for estimating evapotranspiration, soil water content and grain yield of winter wheat in Tensift Al Haouz (Morocco): Application to irrigation management. Agricultural Water Management. 2016;**163**:219-235

[40] Arcand MM, Lynch DH, Voroney RP, van Straaten P. Residues from a buckwheat (Fagopyrum esculentum) green manure crop grown with phosphate rock influence bioavailability of soil phosphorus. Canadian Journal of Soil Science. 2010; **90**(2):257-266

[41] Mollier A, Li H, Shi Y, Ziadi N, Morel C. Soil tillage and P fertilization effects on root distribution and morphology of soybean (Glycin max, L.). In: CSSS/PRSSS Annual Meeting. Canadian Society of Soil Science; 2016. p. 89

[42] Horgan FG, de Freitas TFS, Crisol-Martínez E, Mundaca EA, Bernal CC. Nitrogenous fertilizer reduces resistance but enhances tolerance to the brown planthopper in fast-growing, moderately resistant rice. Insects. 2021, 2021;**12**(11): 989

[43] Tian C, Zhou X, Liu Q, Peng J, Zhang Z, Song H, et al. Increasing yield, quality, and profitability of winter oilseed rape (Brassica napus) under combinations of nutrient levels in fertiliser and planting density. Crop and Pasture Science. 2020;**71**(12):1010-1019

Chapter 8

Microbial and Chemical Contamination of Vegetables in Urban and Peri-Urban Areas of Sub-Sahara Africa

Sanata Traoré, Fassé Samaké, Amadou Hamadoun Babana, Eric Williams Cornelius, Gloria Ladjeh Essilfie, Mavis Agyeiwa Aacheampong and Salimatou Samaké

Abstract

Most farmers in urban and peri-urban areas in West Africa have limited land, so practice farming systems targeted at the production of high-value crops used in urban diets, especially exotic vegetables. Moreover, rapid urban population growth and adverse climate change are causing increased demand for food and water, leading to water scarcity in those cities. The intense pressure of increasing food demand in cities pushes small farmers to depend on untreated wastewater, undecomposed manure, and pesticides for their production, which negatively affect the health of the population. This chapter presents an overview of the identification of pathway and levels of vegetables contamination in SSA and the identification of interventions employed to reduce public health risk. The microbiological and chemical assessment of irrigation water, fertiliser and vegetable samples collected from farms and markets in SSA revealed their contamination with pathogenic bacteria above the recommended standard of WHO and ICMSF. They were also contaminated by heavy metal above the safe limits by FAO/WHO and pesticide residues. The treatment of irrigation water, proper composting of manure and appropriate use of pesticides could be complement disinfection of vegetables before consumption to reduce public health risk.

Keywords: West Africa, vegetables, water scarcity, fertiliser, pesticides

1. Introduction

Agriculture is becoming less beneficial to rural farmers in the Sahelian region of West Africa (WA) because of the population explosion, severity and persistence of climate change, scarcity of irrigation infrastructure, degradation of the environment and slow pace of technology change amongst other factors. These challenges have resulted in a large number of rural farmers migrating to cities in search of social and economic opportunities and its accompanying increase in urban food demand [1, 2]. Most farmers in urban and peri-urban areas have limited land, and so practice farming systems targeted at the production of high-value crops used in urban diets, especially exotic vegetables. These farming systems, known as urban and peri-urban agriculture (UPA), have evolved in different African cities, and in WA involves 20 million people essentially farming for subsistence and/or income generation [3–5]. The contribution of urban agriculture (UA) in strengthening the living conditions of urban Africans is unknown or is seen as a small role in the fight against chronic hunger [6, 7]. The urban population involved in urban UA in Africa is estimated at 50% in Ghana [8, 9], 46% in big cities in Mali [10], 80% of families in Brazzaville, 68% in six Tanzanian cities, 45% in Lusaka, 37% in Maputo, 36% in Ouagadougou and 35% in Yaoundé [11, 12].

In Mali, agriculture contributes about 23% of export earnings and 40% of gross domestic product (GDP) [13], with 5.1% growth rate in 2019 [14]. Vegetable production in Malian cities is usually done on open lands along streams, rivers and roadsides. Due to the inaccessibility of clean irrigation water and the high cost of chemical fertiliser, urban and peri-urban farmers use drainage water and untreated animal waste for production. They are also involved in unselective use of pesticides [15, 16] . These indiscriminate practices negatively affect the health of the population [17–19].

The primary public health problem associated with the consumption of vegetables produced using untreated irrigation water in WA include contamination with pathogenic microorganisms such as *Escherichia coli* [20]. Helminths transmitted to crops through soil are also responsible for some parasitic diseases prevalent in sub-Saharan Africa (SSA) [18, 21, 22]. Studies in West African countries have generally revealed high levels of harmful microorganisms, heavy metals and pesticide residue in irrigation water and vegetables [23–25] far exceed recommended levels set by World Health Organisation (WHO) [26] and International Commission of Microbiological Specifications for Foods (ICMSF) [27]. Many authors [16, 19, 28] also reported high contamination levels of lettuce with pesticide residue and *Salmonella* species in WA. Other biological and chemical contaminants of water for UPA in WA include dye effluents from textile dyeing activities and industrial and domestic waste. The effective wastewater treatment published by WHO [26] is impracticable in most parts of SSA due to high cost. Wastewater treatment for urban vegetable production is therefore currently not a realistic option for WA, and banning the use of untreated irrigation water will also threaten many livelihoods, affect urban vegetable supply, and will therefore be-contrary to poverty alleviation. In order to mitigate these problems, the characteristic of UPA has been reviewed for major cities in SSA countries such as Ghana [29], Burkina Faso [30], Kenya [31], Nigeria [32] etc. The present chapter reviews the identification of pathway and levels of vegetables contamination and also the identification of interventions employed to reduce public health risk without compelling farmers to change their cropping system or water sources.

2. Material and methods

For this study, research was conducted using scientific papers, books, reports and statistical data from WHO to quantify the level of vegetable contamination and determine key points where necessary interventions could be applied to reduce public health risks in SSA. The concept is based on the biological and chemical identification of irrigation waters, fertilisers and vegetables, pre- and post-harvest risk management strategies for pathogens. Microbiological analysis and helminth eggs were done using standard methods of ISO and standard morphological characteristics, respectively. Heavy metals and pesticide residues were quantifying by atomic absorption spectrophotometry and Gas Chromatography, respectively.

3. Contribution of urban and peri-urban agriculture to food security and livelihoods in African cities

Food insecurity is a significant problem in Africa due to rapid urbanisation as urban needy people spend 60–80% of their income on food which, is of low quality and quantity [5, 33]. This food insecurity pushes the rural population to migrate to cities in search of better living conditions. UPA is very beneficial in Africa as it supplies vegetables to urban areas within 30 km radius and this contribute about 70% of their food's needs [34]. In the short term, it increases food security, diversification, food self-sufficiency, earnings, job creation and encourage the use of municipal waste as composts [35, 36].

4. Challenges of vegetables production in SSA

The lack of training of farmers in West Africa on education on innovative agricultural practices and agricultural extension affects vegetable production by causing the yield loss. The yield loss negatively impacts the incomes of farmers and their means of subsistence. Some agricultural practices such as irrigation of vegetable farms with contaminated water, the use of animal or human wastes, pre and post-harvest handling practices, transportation, storage and processing are factors that contribute to bacterial and fungal contamination of vegetables [25, 28, 37–40]. In addition, contaminated water (irrigation water) used to wash vegetable from the farm before they are put on sale are probable factors contributing to parasitic contamination on soil and transmission to vegetables [37].

5. Quality of organic fertilisers and irrigation water used for vegetable production

5.1 Quality of organic fertilisers

One of the good agricultural practices is the use of organic and inorganic fertilisers that increase soil fertility and plant quality [41]. These two fertilisers provide essential food components for the development of the plant [42]. Similar to other African countries, fresh organic manure and compost are the commonly used soil fertility improvement materials in Malian cities [43]. However, the application of organic manures promotes microbiological contamination of soil and vegetables [15, 37, 44, 45].

5.2 Quality of irrigation water used for vegetable production

Rapid urban population growth and adverse climate change in Africa are causing increased demand for food and water, leading to water scarcity in cities. The intense pressure of increasing food demand in cities is pushing small farmers to depend on untreated wastewater for their production [46]. Wastewater, rich in organic matter, is available, freely (because it comes from stagnant streams or lakes), and is beneficial for soil fertility [47]. However, wastewater has a high level of contamination with microbial pathogens and chemicals. African cities do not have adequate sewage facilities, pushing the population to dump waste directly into waterways [48]. Small urban farmers find it difficult, often impossible to get good quality water (visually clear water), and eventually use this water (of poor quality) for their production. Studies in WA have shown high levels of contamination of irrigation water with foodborne pathogens (faecal coliforms) above WHO [26] recommended standard [49–51]. Akinde et al. [28] found total coliforms (1.27 10^4 CFU/ml), *Klebsiella pneumoniae* (27.9%), *Pseudomonas aeruginosa* (26.2%), *E. coli* (8.2%), *Citrobacter*, *Enterobacter* and *Salmonella* spp.) in Nigeria. These pathogens are from the faeces of animals or humans or domestic and industrial waste [52, 53]. They cause public health problems because they can be a source of disease amongst consumers of these vegetables (especially those consumed raw) [54]. These diseases can be infectious and are amongst the leading causes of death in the world [55]. In developing countries, 80% of death are from water-related diseases [56]. Several epidemiological studies have shown links between the use of wastewater in agriculture and the risk of infections (viral, protozoan, intestinal and bacterial) amongst farmers, their families, and consumers [38, 39, 57]. Several studies conducted in WA over the last decade have linked vegetable contaminations with helminth eggs such as *Ascaris lumbricoides*, *Hymenolepis diminuta, Trichuris trichiura, Fasciola hepatica* and *Strongyloides larvae* to use of untreated irrigation water [49, 51, 58]. At-risk groups are farmers, consumers, children, especially the general population living in the vicinity of an unsuitable irrigated water field [59]. Amoah et al. [18] found higher prevalence of soil-transmitted helminth (6–16.10^3 eggs/L) in wastewater irrigation areas in developing countries.

Heavy metals pollution in agricultural soils and water are mainly from anthropogenic sources such as fertilisers, waste disposal, mining, smelting, sewage sludge, urban effluent and vehicle exhausts [60]. Heavy metals such as lead, cadmium and mercury, have been found in irrigation water in many countries in WA and their concentration varies depending on the region of crop as well the type of metals [24, 51, 61, 62]. The source of heavy metal contamination in water in Mali include industrial and domestic waste and dye effluents from textile dyeing activities (which are poured directly into rivers or drains (that flow directly into rivers) without receiving suitable treatment).

Many pesticides are not easily degradable, some persist in the soil, penetrate surface and groundwater and contaminate the environment. They often penetrate the organism according to their chemical properties and cause bioaccumulation. The duration of pesticides in the soil varies from a few hours to several years in the case of organochlorine pesticides [63]. The aerial drift of pesticides, the wind erosion of treated soil and the volatilisation of applications in terrestrial environments are mostly responsible for pesticide contamination of aquatic ecosystems [19]. Farmers in Africa generally do not follow safety measures in the formulation and application of pesticides and ignore the environmental impacts [64, 65]. In general, pesticides found in water in Africa are organochlorines [66].

6. Quality of fresh vegetables in farms and markets in SSA

Enteric bacteria can contaminate vegetables during cultivation, harvesting, transportation and processing. A study by Alemu et al. [67] in Ethiopia showed the contamination of 48.7% of the vegetables sold (cabbage predominated with 71.9%) in Arba Minch Town (Southern Ethiopia) by bacteria. The most frequently detected contaminant amongst vegetables was *E. coli* (31.4%). Microbiological analysis of endives, tomatoes, and ready-to-eat raw produce sold on the market in Côte d'Ivoire revealed a high level of contamination by faecal coliforms, *Enterococcus* and *Pseudomonas*. Washing of vegetables reduced microbial load [68]. Cabbage, lettuce, and scallions sold in Abura and Kotokuraba markets in Ghana were found to be contaminated by *E. coli*, *Enterobacter* spp., *Klebsiella* spp., *Salmonella* spp., *Serratia marcescens*, and *Staphylococcus* [40]. The resistance of infectious stages (cysts, oocysts, and eggs) of helminth parasites on irrigated vegetables to chemicals disinfectants, high temperatures and desiccation is responsible for their high prevalence in developing countries [69]. Alemu et al. [67] reported that vegetables sold in markets in Ethiopia contained in a bucket with water and without prior washing had a high risk of being contaminated by parasites. The level of produce contamination depends on people's health habits as well as sanitation in a particular environment, such as the location of market as an example and the water (mainly pipe borne water which contained low parasites) used to wash vegetables [70]. Adamu et al. [71] reported that helminth parasite contamination of vegetables sold in Maiduguri (Nigeria) market was due to irrigation water. Illiteracy and insufficient public health awareness lead residents to indiscriminate defecation resulting in helminth contamination of water bodies and agricultural land. Animal and human waste used as manure to supplement fertilisers and contaminated water used to wash produce such as lettuce and carrots from the farm before they are put on sale also promote parasitic contamination [37]. Lettuce, which is a crop positioned close to the ground, easily become contaminated by geohelminths during heavy rains and floods [70]. Unhygienic harvesting methods, as well as hands contaminated with faeces, and sprinkling of water on vegetables to make them fresh, can also contaminate vegetables, as is the case with tomatoes [72]. Helminth infection in SSA, results in skin irritation, blood and nervous disorders, tumours, infertility problems, breathing difficulties, nausea, vomiting and endocrine disruption amongst others [73, 74]. The risks linked to the consumption of vegetables irrigated with untreated water were described in the study of epidemics. Studies on the risks of typhoid, gastroenteric, cholera, shigellosis, hepatitis and amebiasis due to the consumption of vegetables irrigated with untreated wastewater have been reported [75, 76]. In South Africa, several food-borne disease outbreaks partly attributed to vegetable contamination have been reported particularly for school children [77–79]. Newell et al. [80] reported many cases of food-borne disease outbreaks from 1990 to 2010 in Ghana, Nigeria and Sudan due to the consumption of untreated irrigated vegetables. *Salmonella* spp., *Campylobacter* spp., *E. coli* and *Listeria monocytogenes* were identified as the major bacterial pathogens in the aforementioned outbreaks. In Mali, outbreak of foodborne diseases sometimes occurs, but there are not official records on them.

The content of heavy metals in plants depends on their ability to absorb them selectively and on the characteristics of the soil [81]. Heavy metals can also be absorbed by the aerial parts of plant leaves from dust and atmospheric precipitation in polluted areas, or from the fossil fuels used for heating, traffic density, fertilisers, and protective agents [82, 83]. Vegetables may also be contaminated with trace metals from the water used by farmers to wash vegetables before they are placed on the market [84]. Vegetables grown in urban and peri-urban areas are more likely to be contaminated with high levels of heavy metals. However, the levels of heavy metals in vegetables grown in Ghana were below the recommended dietary thresholds [85]. Plant roots and leaves accumulate high concentration of metals (from soil and irrigation water) relative to other crops and parts of plants [86, 87]. The closer the plant is to the soil, the higher the transfer of heavy metals from the soil to the plant. Ali et al. [88] reported high levels of heavy metal contamination on ready-to-eat lettuce and spinach in Nigeria, which were above the safe limits by FAO/WHO [89]. Some of these metals are toxic to humans, even at low concentrations. Ogunkunle et al. [90] found high concentrations of lead in cabbage and lettuce (1.840 and 1.930 mg/kg respectively) that were above the WHO maximum value or the level authorised by the FAO. Meanwhile the concentrations of cadmium, zinc, copper, cobalt, manganese, and nickel were below the standards recommended by WHO. Aber et al. [84] revealed that cadmium, lead, and mercury were major contaminants in vegetables sold on the streets of Nigeria. Lente et al. [61] reported lead (1.8–3.5 mg/kg), contamination of vegetables irrigated with waste and ground water above the safe limits in Ghana. Taghipour & Mosaferi [91] also found high level of contamination of kurrat, onion, and tomatoes by heavy metals, which was higher than the standard levels. Excess lead and cadmium in food can be responsible for many diseases, especially nerve, bone, kidney and cardiovascular disease that can lead to death [92, 93]. The rapid growth of industries is of concern to the population regarding the potential accumulation of heavy metals in plants, water and soil [94, 95]. Studies have shown that excessive assimilation of heavy metals by plants gives them toxic effects which could enter food chains, biomagnify and be a source of potential threat to the health of the population [96–98].

In controlling pests, smallholders rely heavily on pesticides in the production of vegetables [65]. Bertrand [99] reported the use of 50% herbicides in horticulture in Africa, more than 20 active ingredients are used in Benin and Senegal [101]. According to Kariathi et al. [101], the largest consumers of tomatoes contaminated by pesticides are the most exposed. The intense application of pesticides has also affected the environment, water and air [102, 103]. Fresh produces contain high levels of pesticides compared to other types of food, and their consumption poses a high risk to human health [19, 103, 104]. The most widely used synthetic insecticides active ingredients are organophosphates, carbamates, pyrethroids, and organochlorines and their toxicity can be very high in the short term or the long term [105]. The misuse of pesticides in SSA is due to the high level of illiteracy, lack of training, ignorance, maladjustment to the product and insufficient knowledge of the pests and diseases [106]. The level of toxicity of pesticides varies depending on the toxicity class [100]. Factors such as type of pesticide, location of pesticide deposition, temperature, pH, duration of pesticide sprayed during growth, duration of treatment and nature of the vegetable influence the stability of the pesticide [107]. The health effects of pesticides appearing on the body of users after contamination, depend on the duration and extent of the exposure. These effects can range from mild skin irritation to congenital disabilities, tumours, genetic changes, blood and nervous disorders, endocrine disruption and even coma or death [108, 109]. Pesticides also affect the reproductive, endocrine, and immune systems [110] and symptoms such as headache, stomach and chest pain, skin and eye irritation, breathing difficulties and vomiting amongst farmers in Tanzania have been reported [74]. Better education on handling practices and sometimes stricter controls on the distribution of pesticides is necessary to minimise their effects on the health of sprayers and vegetable consumers.

7. Health risk management strategies

7.1 Pre-harvest risk management measures for pathogens

Risk management strategies in agriculture include biological control, integrated pest management and the use of physical barriers such as greenhouses or plastic tunnels [111]. Guideline procedures have been published by WHO [26] for the management of health risks in wastewater irrigation. These instructions are described in eight methods: wastewater treatment; crop restriction; wastewater application techniques that minimise contamination; withholding periods to allow pathogen die off between last irrigation and consumption; hygienic practices at food markets and food preparation measures; health and hygiene promotion; produce washing, disinfection and cooking; and chemotherapy and immunisation. The methods to identify and quantify the public health risks and decisions by risk managers to determine appropriate control measure strategies are as follows: "(1) quantitative microbial risk assessment (QMRA) to determine the overall reduction of essential pathogens to achieve the health objective for different routes of exposure; (2) determination of the necessary degree of reduction of pathogens to achieve the health objective (3) knowledge of the number of pathogens in untreated wastewater is essential; (4) choice of control measures or intervention methods to achieve a safe level of reduction of pathogens; and (5) establishing the level of monitoring of treatment verification according to the pathogen or an indicator e.g. *E. coli* in the cultures or the final effluent from the treatment plant". Amongst all these methodologies, the implementation of QMRA is perhaps the most important. It provides a scientific basis for comparing the types of risks and solutions. Studies on QMRA in the health field are difficult to adopt due to insufficient information provided by stakeholders and limited resources [112]. Public health and wastewater treatment can be improved through the provision of adequate sanitation and wastewater treatment infrastructure like waste stabilisation ponds. This infrastructure will allow better management of wastewater while improving its quality; evaluate control during and after harvest as well as human exposure [113]. The quality of water used for irrigation can be improved after the first treatment. Primary treatment of wastewater involves different options such as sedimentation or clarification of solid waste within the water; chemically-enhanced sedimentation and up flow anaerobic sludge blanket reactors [114, 115]. Secondary treatment is necessary for complete elimination of biological and chemical contamination. It is possible to do the secondary treatment by various methods, including artificial wetlands, seepage-percolation, and anaerobic upwelling reactors [116]. Hygienic measures can also be applied by farmers to minimise the risk of contamination and the spread of sewage infections. These measures constitute farmers' awareness of the dangers; wearing gloves and boots, a vaccination campaign against diseases transmitted by wastewater; changing irrigation methods, and washing the arms and legs after using wastewater [117]. On-farm risk management measures for pathogens are only possible with the implementation of appropriate practices at the farm level, such as the use of good quality water and appropriate methods for irrigation, wearing gloves and boots by farmers, regular disinfection of work tools after use.

Fertiliser application measures, fundamental amendments, as well as sanitary measures, must be applied to the soil [117]. The fundamental amendments are not to produce edible plants, especially if inorganic contaminants contaminate the soil. The application of limestone is useful for soils contaminated by a high level of sodium. Fertiliser should not be used on acid soils to avoid chemical reactions. In this case, specific management of acidic soil treatment is necessary while selecting crops that do not accumulate hard metals. For calcareous soils, there is no appropriate treatment. Soil treatment time varies depending on the type and degree of contamination [118]. Drip irrigation with treated wastewater and appropriate planting area is also an effective way to reduce pathogen contamination of crops [119]. The type of fertiliser used in urban vegetable production in Africa is dependent on the crop cultivated. Fresh organic manure and compost are commonly used for vegetables production in African countries compared to synthetic fertilisers. Fallow system in SSA, important in agriculture, contributes to improve soil fertility [43, 120].

7.2 Post-harvest risk management measures for pathogens

Measures used for post-harvest risk management of pathogens include improving the harvesting process, cleaning harvested vegetables, good sanitary measures during transportation, selling and disinfecting before consumption. Vegetables should be cut above a certain height of soil (greater than 10 cm) and washed in the field to reduce contamination by pathogens. Transport vehicles must be well equipped for the specific produce and disinfected before and after transport. In this case, the vegetables must be put in hermetically sealed sterile boxes. Vegetables once on the market should be cleaned, put on clean shelves while being covered by clean covers for sale. Wash rim at the market must be changed regularly. In the kitchen, vegetables should be kept in a well cleaned place and disinfected before consumption, especially those consumed raw. Cooking can also result in about 5–6 log reductions in pathogen loads on vegetable consumed uncooked as lettuce [26]. However, there are a combination of food preparation measures that have small reductions of pathogens of leafy vegetables like lettuce consumed uncooked. This whole process helps reduce biological contamination. Chemical contaminants (heavy metals and pesticide residue) are impossible to remove once absorbed [121], hence all potential sources of contaminations must be addressed.

7.2.1 Sanitisers used to decontaminate harvested vegetable produce

Disinfection of vegetables before consumption is necessary to reduce microbial numbers. Several factors affect the effectiveness of disinfection including the type of disinfectant, the susceptibility of pathogens to antimicrobials and their topographic characteristics. For example, the skin of vegetables made of a multi-layer hydrophobic cuticle of cutin, makes their surfaces highly water-repellent [122]. Nonetheless disinfection remains important in reducing microbial contaminations.

7.2.1.1 Tap water

For centuries, tap water has been used to decontaminate vegetables. A survey in Lomé on the use of disinfectants in households revealed that 8% of consumers cleaned vegetables with water [123]. Coulibaly-Kalpy et al. [68] in Côte d'Ivoire showed that washing tomatoes and endives with tap water for 10 min reduced the rate of faecal coliforms, *Enterococci* and *Pseudomonas* by more than 70%. However, Subramanya et al. [124] showed a small reduction (one log CFU/ml) of the bacterial load on the coriander leaves washed with tap water. Traoré et al. [125] also have showed that tap water was not effective in reducing faecal coliforms on lettuce (1.3, 1.6 and 1.9 log CFU/100 g at 5, 10 and 15 minutes, respectively). In addition, Bhilwadikar et al. [122] signalise that tap water was not very effective

(11.1% to 23.7%) in reducing the residue of dimethoate, pirimiphosmethyl and malathion in potatoes. Amoah [126] demonstrated that water is not very effective for removing helminth eggs.

7.2.1.2 Bleach (sodium hypochlorite: NaClO)

Bleach is used as a disinfectant of choice because of its moderate cost, ease of use and safe and environmentally friendly nature [127]. Eighteen percent of families in Lomé used NaClO to disinfect vegetables. Washing tomatoes and endives with NaClO (17.8 mg/l) solution for 10 min reduced the rate of faecal coliforms, *Enterococci* and *Pseudomonas* by more than 95% [68]. Disinfection of lettuces leaves in Mali for 5(0.00855 ppm), 10(0.00570 ppm) and 15 min (0.00285 ppm) in NaClO solution led to 2.5, 2.8 and 3.5 log reduction in faecal coliforms and complete elimination of *Salmonella* spp. [125]. However, the same authors [128] previously obtained complete disinfection of lettuce by using 2.6 mg/l of bleach for 15 min, which did not affect its quality and the chlorine residues on the produce was less than the maximum acceptable value (5 mg/l) in drinking water [129]. Using sodium hypochlorite at higher concentrations may give much better results, but have toxic effect on the human body cells [51].

7.2.1.3 Potassium permanganate ($KMnO_4$)

Potassium permanganate is a powerful oxidant. In Lomé, 22% of consumers washed vegetables with potassium permanganate solution [123]. Traoré et al. [125] reported complete disinfection of *Salmonella* spp. and 3.3 log reductions of faecal coliforms on lettuce with potassium permaganante solution at 170 ppm for 15 min. Cleaning coriander leaves for 10 min in $KMnO_4$ solution (0.1%) eliminated 4 logs CFU of pathogenic bacteria [124]. Subramanya et al. [124] reported 4.1 log reductions of bacterial bioload on raw *Coriandrum sativum* with $KMnO_4$ solution at 0.1% for 10 min. Bhilwadikar et al. [122] revealed the efficacy of $KMnO_4$ (0.001%) in removing residue of carbaryl (93.50%) and methomyl (47.57%) on Chinese cabbage at 15 min. Potassium permanganate at the highest concentration caused purpled discoloration on vegetables (e.g. lettuce leaves) [125].

7.2.1.4 Vinegar

A study by [125] showed the effectiveness of 3.2 log CFU/100 g 0.00285 ppm vinegar solution applied for 15 min in reducing the faecal coliforms present in lettuce. Amoah [126] in Ghana reported that dipping lettuce leaves in vinegar 12,500 ppm for 10 min reduced faecal coliforms populations by 3.5 log units. Disinfecting of vegetables with vinegar have sometime negative sensory effects on vegetables [125, 126].

7.2.1.5 Common salt (NaCl)

Adjrah et al. [123] point out that 22% of consumers used saline water to disinfect vegetables in Lomé. Traoré et al. [125] demonstrated the efficacy of salt in reducing faecal coliforms and eliminate *Salmonella* spp. populations on vegetables. Amoah [126] in Ghana reported that washing lettuce with salt 35 ppm for 10 min reduced faecal coliforms populations by 2.1 log units. High concentrations of salt, however, deteriorate the external quality of some vegetables such as lettuce [126].

7.2.1.6 Electrolysed water (EOW)

It is a powerful bactericide [130, 131]. Studies have shown the use of different types of EOW in disinfection of fresh produce [132–134]. A study by Chen et al. [135] showed the reduction of cyanidine, pelargonidine and 2,2-diphenyl-1-picrylhydrazyle (DPPH) radicals in red cabbage treated with EOW. Many studies demonstrated the effectiveness of EOW against aerobic mesophiles, enterobacteria, mould and yeast contaminating the surface of vegetables [136–138]. The acidic EOW is very effective, but its corrosive and unstable characteristics limit its application [139].

8. Conclusions

This chapter has provided a comprehensive discussion of the sources of vegetables contamination in SSA and health reduction intervention approach that can be employed in preventing vegetable contamination. Many farming practices impacted positively and negatively on the health risk of vegetables produced in SSA, particularly in Mali. Generally, vegetables produced are contaminated by pathogenic bacteria, helminth eggs, heavy metals and pesticides. In addition to irrigation water and soil, possible sources of vegetables contaminations include the use of untreated organic manure. It is important to sensitise and train farmers to improve their knowledge in good agronomic practices in order to reduce microbial and chemical contamination of vegetables in the farm. Some of these practices could be the treatment of irrigation water, proper composting of manure and appropriate use of pesticides. These farm contamination reduction measures could complement disinfection of vegetables before consumption to save the health of vegetable consumers.

Author details

Sanata Traoré[1]*, Fassé Samaké[1,2], Amadou Hamadoun Babana[1], Eric Williams Cornelius[3], Gloria Ladjeh Essilfie[3], Mavis Agyeiwa Aacheampong[3] and Salimatou Samaké[2]

1 Faculty of Sciences and Techniques, University of Sciences, Techniques and Technologies of Bamako, Bamako, Mali

2 Institute of Applied Sciences, University of Sciences, Techniques and Technologies of Bamako, Bamako, Mali

3 Department of Crop Science, University of Ghana, Accra, Ghana

*Address all correspondence to: sanatatraore21@yahoo.fr

References

[1] Henderson JV, Storeygard A, Deichmann U. 50 years of urbanization in Africa: Examining the role of climate change (Report No. WPS6925). 2015. Available from: documents1.worldbank.org/curated/en/861451468192245824/pdf/WPS6925.pdf

[2] FAO. The future of food and agriculture: Trends and challenges. 2017. Available from: www.fao.org/3/a-I6881e.pdf

[3] Drechsel P, Graefe S, Sonou M, Cofie OO. Informal Irrigation in Urban West Africa: An Overview (Report No. 102). Colombo, Sri Lanka: International Water Management Institute; 2006 Available from: documents1.worldbank.org/curated/en/861451468192245824/pdf/WPS6925.pdf

[4] FAO. The place of urban and peri-urban agriculture (UPA) in national food security programmes. 2011. Available from: https://www.fao.org/3/i2177e/i2177e00.pdf

[5] Orsini F, Kahane R, Nono-Womdim R, Gianquinto G. Urban agriculture in the developing world: A review. Agronomy for Sustainable Development. 2013;**33**(4):695-720. DOI: 10.1007/s13593-013-0143-z

[6] Conceição P, Levine S, Lipton M, Warren-Rodríguez A. Toward a food secure future: Ensuring food security for sustainable human development in sub-Saharan Africa. Food Policy. 2016;**60**:1-9. DOI: 10.1016/j.foodpol.2016.02.003

[7] Tibesigwa B, Visser M. Assessing gender inequality in food security among small-holder farm households in urban and rural South Africa. World Development. 2016;**88**:33-49. DOI: 10.1016/j.world-dev.2016.07.008

[8] Obosu-Mensah K. Changes in official attitudes towards urban agriculture in Accra. African Studies Quartely. 2002;**6**(3):19-32

[9] Nchanji EB. Sustainable urban agriculture in Ghana: What governance system works? Sustainability. 2017;**9**:2090. DOI: 10.3390/su9112090

[10] World Bank. The World Bank in Mali. 2015. Available from: www.worldbank.org/en/country/mali/overview

[11] Shackleton CM, Pasquini MW, Drescher AW (1st Ed.). African Indigenous Vegetables in Urban Agriculture Earthscan. 2009. 39p. Available from: https://publications.cta.int/media/publications/downloads/1537_PDF.pdf

[12] Hallett S, Hoagland L, Toner E (1st Ed.). Urban Agriculture: Environmental, economic, and social perspectives. In Janick J. Horticultural Reviews. 2017;**44**(XLIV):65-120. Available from: https://doi.org/10.1002/9781119281269.ch2

[13] Sanogo OM, Doumbia S, Descheemaeker K. Complémentation des bovins laitiers pour l'amélioration de la production de lait et du fumier en milieu paysan dans le cercle de Koutiala. Revue Malienne de Science et de Technologie. 2019;**0**(22):134-143

[14] World Bank. Agriculture, forestry, and fishing, value added (annual % growth). 2019. Available from: https://data.worldbank.org/indicator/NV.AGR.TOTL.KD.ZG

[15] Samaké F, Babana AH, Yaro F, Cissé D, Traoré I, Kanté F, et al. Risques sanitaires liés à la consommation des produits maraîchers cultivés dans la zone urbaine et périurbaine de Bamako. Mali Santé Publique. 2011;**1**(001):27-31

[16] Dia S. Contrôle de la qualité bactériologique et toxicologique de la laitue vendue en commune I du District de Bamako, Mali [Mémoire de Master non publié]. Université des Sciences, des Techniques et des Technologies de Bamako; 2017

[17] Boulanger M, Tual S, Lemarchand C, Guizard AV, Velten M, Marcotullio E, et al. Agricultural exposure and risk of bladder cancer in the agriculture and cancer cohort. International Archives of Occupational and Environmental Health. 2017;**90**(2):169-178. DOI: 10.1007/s00420-016-1182-y

[18] Amoah ID, Adegoke AA, Stenström TA. Soil-transmitted helminth infections associated with wastewater and sludge reuse: A review of current evidence. Tropical Medicine and International Health. 2018;**23**(7):692-703. DOI: 10.1111/tmi.13076

[19] Abagale A, Atiemo S, Abagale FK, Ampofo A, Amoah CY, Aguree S, et al. Pesticide residues detected in selected crops, fish and soil from irrigation sites in the upper east region of Ghana. Advanced Journal of Chemistry-Section A. 2020;**3**(2):221-236. DOI: 10.33945/SAMI/aJCA.2020.2.10

[20] Amoah P, Keraita B, Drechsel P, Abaidoo R, Konradsen F, Akple M. Low cost options for health risk reduction where crops are irrigated with polluted water in West Africa (Report No. 141). Colombo, Sri Lanka: International Water Management Institute. 2011. Available from: https://www.iwmi.cgiar.org/publications/IWMI_Research_Reports/PDF/PUB141/RR141.pdf

[21] Akhlaghi L, Mafi M, Oormazdi H, Reza MA, Shirbazou S, Tabatabaie F. Survey of parasitic infections of consumed vegetables in Ahwaz city. Annals of Biological Research. 2013;**4**(8):22-26

[22] WHO. Soil-Transmitted Helminth Infections. 2020. Available from: https://www.who.int/news-room/fact-sheets/detail/soil-transmitted-helminth-infections

[23] Antwi-Agyei P, Cairncross S, Peasey A, Price V, Bruce J, Baker K, et al. A farm to fork risk assessment for the use of wastewater in agriculture in Accra, Ghana. PLoS One. 2015;**10**(11):e0142346. DOI: 10.1371/journal.pone.0142346

[24] Salawu K, Barau MM, Mohammed D, Mikailu DA, Abdullahi BH, Uroko RI. Determination of some selected heavy metals in spinach and irrigated water from Samaru area within Gusau Metropolis in Zamfara State, Nigeria. Journal of Toxicology and Environmental Health Sciences. 2015;7(8):76-80. DOI: 10.5897/JTEHS2015.0339

[25] Dao J, Stenchly K, Traoré O, Amoah P, Buerkert A. Effects of water quality and post-harvest handling on microbiological contamination of lettuce at urban and peri-urban locations of Ouagadougou. Burkina Faso. Foods. 2018;7(12):206-218. DOI: 10.3390/foods7120206

[26] WHO guidelines for the safe use of wastewater, excreta and greywater use in agriculture; 2006. Available from: whqlibdoc.who.int/publications/2006/9241546832_eng.pdf?ua=1

[27] ICMSF. The 2nd Edition of Microorganisms in Foods 7. Springer; 2018;487. Available from: https://link.springer.com/book/10.1007/978-3-319-684604#authorsandaffiliationsbook

[28] Akinde SB, Sunday AA, Adeyemi FM, Fakayode IB, Oluwajide OO, Adebunmi AA, et al. Microbes in irrigation water and fresh vegetables: Potential pathogenic bacteria assessment and implications for food

safety. Journal of ABSA International. 2016;**21**(2):89-97. DOI: 10.1177/1535676016652231

[29] Obuobie E, Hope L. Characteristics of urban vegetable farmers and gender issues. In: Drechsel P, Keraita B, Keraita (2nd editors). Irrigated Urban Vegetable Production in Ghana: Characteristics, Benefits and Risk Mitigation. Colombo, Sri Lanka: International Water Management Institute; 2014. pp. 28-37. Available from: https://www.iwmi.cgiar.org/Publications/Books/PDF/irrigated_urban_vegetable_production_in_ghana.pdf

[30] Kamga RT, Some S, Tenkouano A, Issaka YB, Ndoye O. Assessment of traditional African vegetable production in Burkina Faso. Journal of Agricultural Extension and Rural Development. 2016;**8**(8):141-150. DOI: 10.5897/JAERD2016.0788

[31] Ochilo WN, Nyamasyo GN, Kilalo D, Otieno W, Otipa M, Chege F, et al. Characteristics and production constraints of smallholder tomato production in Kenya. Scientific African. 2019;**2**:e00014. DOI: 10.1016/j.sciaf.2018.e00014

[32] Taiwo OJ. Analysis of urban agricultural characteristics along land use gradient in Lagos state, Nigeria. Ife Research Publications in Geography. 2016;**14**:1-16

[33] Magnusson U, Bergman KF, Katunguka RE. Introduction to urban and peri-urban agriculture for food security. In: Magnusson U, Bergman KF, editors. Urban and Peri-Urban Agriculture for Food Security in Low-Income Countries: Challenges and Knowledge Gaps. SLU Global; 2014. pp. 4-5. Available from: https://www.siani.se/wp-content/uploads/files/document/slu-global-report-2014-4-urban-and-peri-urban-agriculture-for-food-security-webb.pdf

[34] Karg H, Drechsel P. Atlas of West African Urban Food Systems: Examples from Ghana and Burkina Faso. International Water Management Institute; 2018. Available from: https://ruaf.org/assets/2019/11/atlas_of_west_african_urban_food_systems_examples_from_ghana_and_burkina_faso.pdf

[35] Zezza A, Tasciotti L. Urban agriculture, poverty and food security: Empirical evidence from a sample of developing countries. Food Policy. 2010;**35**(4):265-273. DOI: 10.1016/j.foodpol.2010.04.007

[36] Arku G, Mkandawire P, Aguda N, Kuuire V. Africa's Quest for Food Security: What Is the Role of Urban Agriculture? (Paper No. 19). The African Capacity Building Foundation; 2012. Available from: allafrica.com/download/resource/main/main/idatcs/00060351:9f53711217e8321b8aff8efa6bc9dadf.pdf

[37] Jiang X, Chen Z, Dharmasena M. The role of animal manure in the contamination of fresh food. In: Sofos J, editor. Advances in Microbial Food Safety. Woodhead Publishing Series in Food Science, Technology and Nutrition; 2015. pp. 312-350. Available from: https://reader.elsevier.com/reader/sd/pii/B978178242107850013X?token=1BA45FCEEC1E527043D931B12C68AA4125CD39D55ECC9C29A3CD554E070687B62493D9E66F1C68B744FA082128A415EA&originRegion=eu-west-1&originCreation=20220919161252

[38] Adegoke AA, Amoah ID, Stenström TA, Verbyla ME, Mihelcic JR. Epidemiological evidence and health risks associated with agricultural reuse of partially treated and untreated wastewater: A review. Frontiers in Public Health. 2018;**6**:337-357. DOI: 10.3389/fpubh.2018.00337

[39] Bougnom BP, Zongo C, McNally A, Ricci V, Etoac FX, Thiele-Bruhn S, et al.

Wastewater used for urban agriculture in West Africa as a reservoir for antibacterial resistance dissemination. International Journal of Environmental Research and Public Health. 2019;**168**:14-24. DOI: 10.1016/j.envres. 2018.09.022

[40] Yafetto L, Ekloh E, Sarsah B, Amenumey EK, Adator EH. Microbiological contamination of some fresh leafy vegetables sold in Cape Coast, Ghana. Ghana Journal of Science. 2019;**60**(2):11-23. DOI: 10.4314/gjs.v60i2.2

[41] Mofunanya AAJ, Ebigwai JK, Bello OS, Egbe AO. Comparative study of the effects of organic and inorganic fertilizer on nutritional composition of *Amaranthus spinosus* L. Asian Journal of Plant Sciences. 2015;**14**(1):34-39

[42] Aderinoye-Abdulwahab SA, Salami ST. Assessment of organic fertilizer usage by vegetable farmers in Asa local government area of Kwara state, Nigeria. Agrosearch. 2017;**17**(1):101-114

[43] Farnwortha CR, Stirlingb C, Sapkotac TB, Jatc ML, Misikodand M, Attwood S. Gender and inorganic nitrogen: What are the implications of moving towards a more balanced use of nitrogen fertilizer in the tropics? International Journal of Agricultural Sustainability. 2017;**15**(2):136-152. DOI: 10.1080/14735903.2017.1295343

[44] ANSSA. Risques sanitaires liés à la consommation des produits maraîchers cultivés dans la zone urbaine et périurbaine de Bamako. (Rapport No. 002007/ANSSA). Bamako; 2007

[45] Khan MN, Mobin M, Abbas ZK, Alamri SA. Fertilizers and their contaminants in soils, surface and groundwater. In: DellaSala DA, Goldstein MI, editors. The Encyclopedia of the Anthropocene. Elsevier Incorporation; 2018. pp. 225-240. Available from: https://doi.org/10.1016/B978-0-12-809665-9.09888-8

[46] Islam SMF, Karim Z. World's demand for food and water: The consequences of climate change. In: MHDA. Farahani V, Vatanpour AH. Taheri editors, Desalination: Challenges and opportunities. 2019. (pp. 57-84). p. 57. IntechOpen. DOI: 10.5772/intecopen.85919

[47] Kominko H, Gorazda K, Wzorek Z. The possibility of organo-mineral fertilizer production from sewage sludge. Waste Biomass Valor. 2017;**8**:1781-1791. DOI: 10.1007/s12649-016-9805-9

[48] Saghir J, Santoro J. Urbanization in Sub-Saharan Africa: Meeting Challenges by Bridging Stakeholders (Report 2018). Center for Strategic & International Studies; 2018. Available from: public/publication/180411_Saghir_UrbanizationAfrica_Web.pdf

[49] Drechsel P, Keraita B editors. Irrigated urban vegetable production in Ghana: Characteristics, benefits and risk mitigation (2nd ed.). 2014; 249p. Colombo, Sri Lanka: International Water Management. Available from: https://www.iwmi.cgiar.org/Publications/Books/PDF/irrigated_urban_vegetable_production_in_ghana.pdf

[50] Ja'afaru MI, Ewansiha JU, Dahiru A, Chimbekujwo KI. Microbiological, physicochemical and heavy metals assessments of soils and selected vegetables grown on rumde-doubeli irrigated farmland in Yola Nigeria. Tanzania Journal of Science. 2020;**46**(3):684-699. DOI: 10.1007/s13201-017-0598-1

[51] Traoré S. Microbial and Chemical Contamination of Irrigated Lettuce

(*Lactuca sativa* L.) in some Urban and Peri-Urban Areas in Mali. [Unpublished Doctoral Dissertation]. University of Ghana; 2020

[52] Ajonina C, Buzie C, Rubiandini RH, Otterpohl R. Microbial pathogens in wastewater treatment plants (WWTP) in Hamburg. Journal of Toxicology and Environmental Health. 2015;**78**(6):381-387. DOI: 10.1080/15287394.2014.989626

[53] Penakalapati G, Swarthout J, Delahoy MJ, McAliley L, Wodnik B, Levy K, et al. Exposure to animal feces and human health: A systematic proposed research priorities. Environmental Science and Technologies. 2017;**51**:11537-11552. DOI: 10.1021/acs.est.7b02811

[54] Alegbeleye OO, Singleton I, Anderson S, Sant'Ana AS. Sources and contamination routes of microbial pathogens to fresh produce during field cultivation: A review. Food Microbiology. 2018;**73**:177-208. DOI: 10.1016/j.fm.2018.01.003

[55] Rather IA, Koh WY, Paek WK, Lim J. The sources of chemical contaminants in food and their health implications. Frontiers in Pharmacology. 2017;**8**:830-838. DOI: 10.3389/fphar.2017.00830

[56] Daud MK, Nafees M, Ali S, Rizwan M, Bajwa RA, Shakoor MB, et al. Drinking water quality status and contamination in Pakistan. BioMed Research International. 2017;**2017**(2017):e7908183. DOI: 10.1155/2017/7908183

[57] Shah HA, Huxley P, Elmes J, Murray KA. Agricultural land-uses consistently exacerbate infectious disease risks in Southeast Asia. Nature Communications. 2019;**2019**(10):4299. DOI: 10.1038/s41467-019-12333-z

[58] Cobbina SJ, Kotochi MC, Korese JK, Akrong MO. Microbial contamination in vegetables the farm gate due to irrigation with wastewater in the tamale Metropolis of northern Ghana. Journal of Environmental Protection. 2013;**4**(7):676-682. DOI: 10.4236/jep.2013.47078

[59] Kpoda NW, Oueda A, Some YSC, Cissé G, Maiga AH, Kabre GB. Physicochemical and parasitological quality of vegetables irrigation water in Ouagadougou city, Bukina Faso. African Journal of Microbiology Research. 2015;**9**(5):307-317. Available from: https://www.academicjournals.org/AJMR

[60] Latheef S, Soundhirarajan K. Heavy metal contamination in irrigation water and its effects on plants. International Research Journal of Engineering and Technology. 2018;**5**(5):3704-3710. Available from: https://www.irjet.net/archives/V5/i5/IRJET-V5I5794.pdf

[61] Lente I, Ofosu-Anim J, Brimah AK, Atiemo S. Heavy metal pollution of vegetable crops irrigated with wastewater in Accra, Ghana. West African Journal of Applied Ecology. 2014;**22**(1):41-58

[62] Rouamba SS, Guira F, Nikièma F, Sawadogo A, Kabré E, Sangaré L, et al. Lead and cadmium contamination level in irrigation water and lettuce (*Lactuca sativa* L) from market gardening sites of Ouagadougou, Burkina Faso. International Research Journal of Public and Environmental Health. 2021;**8**(3):208-214. DOI: 10.15739/irjpeh.21.020

[63] Lushchak VI, Matviishyn TM, Husak VV, Storey JM, Storey KB. Pesticide toxicity: A mechanistic approach. EXCLI Journal. 2018;**17**:101-1136. DOI: 10.17179/excli2018-1710

[64] Teklu BM, Adriaanse P, Horst MM, Deneer JW, Van den Brink P. Surface water risk assessment of pesticides in Ethiopia. Science of the Total

Environment. 2015;**508**:566-574. DOI: 10.1016/j.scitotenv.2014.11.049

[65] Mengistie BT, Mol AJ, Oosterveer P. Pesticide use practices among smallholder vegetable farmers in Ethiopian central Rift Valley. Environment, Development and Sustainability. 2017;**19**(1):301-324. DOI: 10.1007/s10668-015-9728-9

[66] Montagner CC, de Medeiros JF, de Vilhena AE, Vizioli BC, Nıvea CG, Munin NCG. Pesticides in aquatic matrices in developing countries: What do we know so far? In: Dalu T, Tavengwa NT, editors. Emerging. Freshwater Pollutants. Elsevier; 2022. pp. 203-226. DOI: 10.1016/B978-0-12-822850-0.00003-x

[67] Alemu G, Nega M, Alemu M. Parasitic contamination of fruits and vegetables collected from local markets of Bahir Dar city, Northwest Ethiopia. Research and Reports in Tropical Medicine. 2020;**11**:17-25. DOI: 10.2147/RRTM.S244737

[68] Coulibaly-Kalpy J, Agbo EA, Dadie TA, Dosso M. Microbiological quality of raw vegetables and ready to eat products sold in Abidjan (Côte d'Ivoire) markets. African Journal of Microbiology Research. 2017;**11**(5): 204-210. DOI: 10.5897/AJMR2016.8427

[69] Zewdneh T, Dawit K. Parasitological contamination of wastewater irrigated and raw manure fertilized vegetables in Mekelle city and its suburb, Tigray, Ethiopia. Momona Ethiopian Journal of Science. 2012;**4**(1):77-89. DOI: 10.4314/mejs.v4i1.74058

[70] Amaechi EC, Ohaeri CC, Ukpai OM, Adegbite RA. Prevalence of parasitic contamination of salad vegetables in Ilorin, north central, Nigeria. Momona Ethiopian Journal of Science. 2016;**8**(2):136-145. DOI: 10.4314/mejs.v8i2.3

[71] Adamu NB, Adamu JY, Mohammed D. Prevalence of helminth parasites found on vegetables sold in Maiduguri, Northeastern Nigeria. Food Control. 2012;**25**:23-26. DOI: 10.1016/j.foodcont.2011.10.016

[72] Shenge KC, Whong CMZ, Yakubu LL, Omolehin RA, Erbaugh MJ, Sally A, et al. Contamination of tomatoes with coliforms and *Escherichia coli* on farms and in markets of Northwest Nigeria. Journal of Food Protection. 2015;**78**(1):57-64. DOI: 10.4315/0362-028X.JFP-14-265

[73] Hotz C, Loechl C, Lubowa A, Tumwine JK, Ndeezi G, Nandutu Masawi A, et al. Introduction of *β*-carotene-rich orange sweet potato in rural Uganda resulted in increased vitamin a intake among children and women and improved vitamin a status among children. Journal of Nutrition. 2012;**142**:1871-1880. DOI: 10.3945/jn.111.151829

[74] Mdegela RH, Mosha RD, Ngowi HA, Nonga H. Environmental and health impacts associated with usage of agrochemicals in Mindu dam catchment area, Morogoro, Tanzania. Huria Journal of the Open University of Tanzania. 2013;**15**(1)

[75] Jiménez B, Navarro I. Wastewater use in agriculture: Public health considerations. In: Jorgensen S, editor. Encyplopeadia on the Environmental Management. CRC Press; 2009. pp. 2694-2708

[76] Jaramillo MF, Restrepo I. Wastewater reuse in agriculture: A review about its limitations and benefits. Sustainability. 2017;**9**:1734-1753. DOI: 10.3390/su9101734

[77] Niehaus AJ, Apalata T, Coovadia YM, Smith AM, Moodley P. An outbreak

of foodborne salmonellosis in rural KwaZulu-Natal, South Africa. Foodborne Pathogens and Disease. 2011;**8**(6):693-697. DOI: 10.1089/fpd.2010.0749

[78] NICD. Food and water-borne diseases. Communicable Diseases Communiqué. 2014;**13**(2):1-13

[79] NICD. Food and water-borne diseases. Communicable Diseases Communiqué. 2014;**3**(11):1-10

[80] Newell DG, Koopmans M, Verhoef L, Duizer E, Aidara-Kane A, Sprong H, et al. Food-borne diseases — The challenges of 20 years ago still persist while new ones continue to emerge. International Journal of Food Microbiology. 2010;**139**:S3-S15. DOI: 10.1016/j.ijfoodmicro.2010.01.021

[81] Morkunas I, Wozniak A, Mai VC, Sobkowiak R, Jeandet P. The role of heavy metals in plant response to biotic stress. Molecules. 2018;**23**(9):e2320. DOI: 10.3390/molecules23092320

[82] Weyens N, Thijs S, Witters RPN, Przybysz A, Espenshade J, Gawronska H, et al. The role of plant–microbe interactions and their exploitation for phytoremediation of air pollutants. International Journal of Molecular Science. 2015;**16**(10):25576-25604. DOI: 10.3390/ijms161025576

[83] Wei X, Lyu S, Yu Y, Wang Z, Liu H, Pan D, et al. Phylloremediation of air pollutants: Exploiting the potential of plant leaves and leaf-associated microbes. Frontiers in Plant Science. 2017;**8**:1318. DOI: 10.3389/fpls.2017.01318

[84] Aber H, Mulindwa J, Lung'aho M, Nyakundi F, Wambui J, Jager M, et al. Hazard analysis and critical control point plan for hazards in Ugandan amaranth vegetable value chain. African Journal of Food Agriculture, Nutrition and Development. 2019;**19**(2):14458-14482. DOI: 10.18697/ajfand.85.17425

[85] FAO/WHO. Evaluations of the Joint FAO/WHO Expert Committee on Food Additives (JECFA). 2014. Available from: apps.who.int/ipsc/database/evaluations/chemical.aspx?chemID=3511

[86] Pandey R, Shubhashish K, Pandey J. Dietary intake of pollutant aerosols via vegetables influenced by atmospheric deposition and wastewater irrigation. Ecotoxicology and Environmental Safety. 2012;**76**:200-208. DOI: 10.1016/j.ecoenv.2011.10.004

[87] Traoré S, Samaké F, Babana AH. Evaluation de la charge microbiologique et physicochimique des légumes issus des sites de grande production maraichère du Mali [Paper presentation]. Bamako, Mali: Malian Symposium for Applied Science; 2018

[88] Ali ZN, Abdulkadir FM, Imam MM. Determination of some heavy metals in spinach and lettuce from selected markets in Kaduna metropolis. Nigerian Journal of Chemical Research. 2012;**17**(2012):23-29. DOI: 10.4314/NJCR.V17I1

[89] FAO/WHO. Report of the 14th session of the codex committee on contaminants in foods [Paper presentation]. Forty-fourth session, joint FAO/WHO food standards programme codex alimentarus commission. 2021. Available from: www.fao.org/fao-who-codexalimentarius/sh-proxy/fr/?lnk=1&url=https%253A%252F%252Fworkspace.fao.org%252Fsites%252Fcodex%252FMeetings%252FCX-735-14%252FREPORT%252FFinalReport%252FREP21_CFe.pdf

[90] Ogunkunle ATJ, Bello OS, Ojofeitimi OS. Determination of heavy metal contamination of street-vended fruits and vegetables in Lagos state,

Nigeria. International Food Research Journal. 2014;**21**(5):1725-1730

[91] Taghipour H, Mosaferi M. Heavy metals in the vegetables collected from production sites. Journals tbzmed. 2013;**3**(2):185-193. DOI: 10.5681/hpp.2013.022

[92] Rahimzadeh MR, Rahimzadeh MR, Kazemi S, Moghadamnia AA. Cadmium toxicity and treatment: An update. Caspian Journal of Internal Medicine. 2017;**8**(3):135-145. DOI: 10.22088/cjim.8.3.135

[93] Rai PK, Lee SS, Zhang M, Tsang YF, Kim KH. Heavy metals in food crops: Health risks, fate, mechanisms, and management. Environment International. 2019;**125**:365-385. DOI: 10.1016/j.envint.2019.01.067

[94] Vongdala N, Tran HD, Xuan TD, Teschke R, Khanh TD. Heavy metal accumulation in water, soil, and plants of municipal solid waste landfill in Vientiane, Laos. International Journal of Environmental Research and Public Health. 2018;**16**:22. DOI: 10.3390/ijerph16010022

[95] Esmaeilzadeh M, Jaafari J, Mohammadi AA, Panahandeh M, Javid A, Javan S. Investigation of the extent of contamination of heavy metals in agricultural soil using statistical analyses and contamination indices. Human and Ecological Risk Assessment: An International Journal. 2019;**25**(5):1125-1136. DOI: 10.1080/10807039.2018.1460798

[96] Patra RC, Rautray AK, Swarup D. Oxidative stress in lead and cadmium toxicity and its amelioration. Veterinary Medicine International. 2011;**2011**:e457327. DOI: 10.4061/2011/457327

[97] Ali H, Khan E, Ilahi I. Environmental chemistry and ecotoxicology of hazardous heavy metals: Environmental persistence, toxicity, and bioaccumulation. Journal of Chemistry. 2019. DOI: 10.1155/2019/6730305

[98] Sugiyama M. Role of cellular antioxidants in metal-induced damage [published in 1994 and was accessed in 2019]. Cell Biology and Toxicology. 2019;**10**:1-22. DOI: 10.1007/BF00757183

[99] Bertrand PG. Uses and misuses of agricultural pesticides in Africa: Neglected public health threats for workers and population. In: Larramendy M, Soloneski S, editors. Pesticides Use and Misuse and Their Impact in the Environment. London: IntechOpen; 2019. pp. 97-110

[100] Rodenburg J, Johnson JM, Dieng I, Senthilkumar K, Vandamme E, Cyriaque A, et al. Status quo of chemical weed control in rice in sub-Saharan Africa. Food Security. 2019;**11**:69-92. DOI: 10.1007/s12571-018-0878-0

[101] Kariathi V, Kassim N, Kimanya M. Risk of exposures of pesticide residues from tomato in Tanzania: Review. African Journal of Food Science. 2017;**11**(8):255-262. DOI: 10.5897/AJFS2016.1527

[102] Mekonen S, Ambelu A, Spanoghe P. Pesticide residue evaluation in major staple food items of Ethiopia using the Quechers method: A case study from the Jimma zone. Journal of Environmental, Toxicology and Chemestry. 2014;**33**(6):1294-1302. DOI: 10.1002/etc.2554

[103] Wolde M, Abirdew S. The significance and implications of pesticide residue on fruits and vegetables in Ethiopia: An overview. American Journal of Bioscience and Bioengineering. 2019;**7**(6):71-81. DOI: 10.11648/j.bio.20190706.11

[104] Pogăcean MO, Hlihor RM, Preda C, Gavrilescu M, M. Humans in the environment comparative analysis and assessment of pesticide residues from field-grown tomatoes. European Journal of Science and Theology. 2013;**9**(6):79-94

[105] Damalas CA, Koutroubas SD. Farmers' exposure to pesticides: Toxicity types and ways of prevention. Toxics. 2016;**2016**(4):1-10. DOI: 10.3390/toxics4010001

[106] Dzisah JS, Sottie CA. Special issue on good governance and sustainable development goals. Ghana Social Science Journal. 2016, 2016;**13**(2):e274

[107] Lopez-Fernandez O, Rial-Otero R, Simal-Gandara J. Factors ´ governing the removal of mancozeb residues from lettuces with washing solutions. Food Control. 2013;**34**(2):530-538. DOI: 10.1016/j.foodcont.2013.05.022

[108] Wang N, Li HB, Long JL, Cai C, Dai JL, Zhang J, Wang RQ. Contamination, Source, and Input Route of Polycyclic Aromatic Hydrocarbons in Historic. 2012

[109] Yegambaram M, Manivannan B, Beach TG, Rolf U, Halden RU. Role of environmental contaminants in the etiology of alzheimer's disease: A review. Current Alzheimer Research. 2015;**12**(2):116-146. DOI: 10.2174/1567205012666150204121719

[110] Al-Waili N, Salom K, Al-Ghamdi A, Ansari MJ. Antibiotic, pesticide, and microbial contaminants of honey: Human health hazards. The Science World Journal. 2012:e930849. DOI: 10.1100/2012/930849

[111] Weintraub PG, Recht E, Mondaca LL, Harari AR, Diaz BM, Bennison J. Arthropod pest management in organic vegetable greenhouses. Journal of Integrated Pest Management. 2017;**8**(1):29. DOI: 10.1093/jipm/pmx021

[112] Jiménez B, Navarro I. Wastewater use in agriculture: Public health considerations. In: Jorgensen S, editor. Encyplopeadia on the Environmental Management. CRC Press; 2012. pp. 3315-3329

[113] Edokpayi JN, Odiyo JO, Durowoju OS. Impact of wastewater on surface water quality in developing countries: A case study of South Africa. In: Tutu H, editor. Water Quality. London: Intechopen; 2017. pp. 401-416. DOI: 10.5772/66561

[114] BLOGANICA. Primary, Secondary, and tertiary wastewater treatment: How do they work? ORGANICA. 2017. Available from: https://www.organicawater.com/primary-secondary-tertiary-wastewater-treatment-work/

[115] Oakley S. Preliminary treatment and primary settling. In: Rose JB, Jiménez-Cisneros B, editors. Water and Sanitation for the 21st Century: Health and Microbiological Aspects of Excreta and Wastewater Management (Global Water Pathogen Project). E. Lansing, MI: UNESCO; 2018. p. 21

[116] Skousen J, Zipper CE, Rose A, Ziemkiewicz PF, Nairn RM, Donald LM, et al. Review of passive systems for acid mine drainage treatment. Mine Water and Environment. 2016;**36**:133-153. DOI: 10.1007/s10230-016-0417-1

[117] Elgallal ME. Development of an approach for the evaluation of wastewater reuse options for arid and semi-arid area [Doctoral dissertation, University of Leeds, United Kingdom]. 2017. Available from: core.ac.uk/download/pdf/83935037.pdf

[118] Ashraf MA, Maah MJ, Yusoff I. Soil contamination, risk assessment and

remediation. In: Hernandez-Soriano MC, editor. Environmental Risk Assessment of Soil Contamination. London: IntechOpen; 2014. pp. 3-56

[119] Khalid S, Shahid M, Natasha BI, Sarwar T, Shah AH, Niazi NK. A review of environmental contamination and health risk assessment of wastewater use for crop irrigation with a focus on low and high-income countries. International Journal of Environmental Research and Public Health. 2018;**15**(5):895-931. DOI: 10.3390/ijerph15050895

[120] Partey ST, Zougmore RB, Ouedraogo M, Thevathasan NV. Why promote improved fallows as a climate-smart agroforestry technology in sub-Saharan Africa? Sustainability MDPI. 2017;**9**:1-12. DOI: 10.3390/su9111887

[121] Keraita B, Silverman A, Amoah P, Asem-Hiablie S. Quality of irrigation water used for urban vegetable production. In: Drechsel P, Keraita B, editors. Irrigated Urban Vegetable Production in Ghana: Characteristics, Benefits and Risk Mitigation. International Water Management Institute; 2014. pp. 62-73

[122] Bhilwadikar T, Pounraj S, Manivannan S, Rastogi NK, Negi PS. Decontamination of microorganisms and pesticides from fresh fruits and vegetables: A comprehensive review from common household processes to modern techniques. Comprehensive Reviews in Food Science and Food Safety. 2019;**18**:36. DOI: 10.1111/1541-4337.12453

[123] Adjrah Y, Karou DS, Djéri B, Anani K, Soncy K, Ameyapoh Y, et al. Hygienic quality of commonly consumed vegetables, and perception about disinfecting agents in Lomé. International Food Research Journal. 2011;**18**(4):1499-1503

[124] Subramanya SH, Pai V, Bairy I, Nayak N, Gokhale S, Sathian B. Potassium permanganate cleansing is an effective sanitary method for the reduction of bacterial bioload on raw *Coriandrum sativum*. BioMed Central Research Notes. 2018;**11**:e124. DOI: 10.1186/s13104-018-3233-9

[125] Traoré S, Cornelius EW, Samaké F, Essilfie G, Babana AH, Bengaly M, et al. Efficiency of common washing treatments in reducing microbial levels on lettuce in Mali, west African. Journal of Applied Ecology. 2020;**28**(2):64-74

[126] Amoah P, Drechsel P, Abaidoo RC, Klutse A. Effectiveness of common and improved sanitary washing methods in selected cities of West Africa for the reduction of coliform bacteria and helminth eggs on vegetables. Tropical Medicine and International Health. 2007;**12**(2):40-50. DOI: 10.1111/j.1365-3156.2007.01940.x

[127] Gomez-Lopez VM, Gil MI, Allende A. A novel electrochemical device as a disinfection system to maintain water quality during washing of ready to eat fresh produce. Food Control. 2017;**71**:242-247. DOI: 10.1016/j.foodcont.2016.07.001

[128] Traoré S, Samaké F, Babana AH, Traoré D, Sanogo Y, Maiga K. Evaluation of the effectiveness of bleach on microbial populations of lettuce. Scientific Journal of Microbiology. 2013;**2**(9):166-173. DOI: 10.14196/sjm.v2i9.950

[129] WHO. Progress on drinking water, sanitation and hygiene. Update and SDG Baselines (Progress report). 2017. Available from: apps.who.int/iris/bitstream/handle/10665/258617/9789241512893-eng.pdf?sequence=1133

[130] Rahman SME, Khan I, Oh DH. Electrolyzed water as a novel sanitizer

in the food industry: Current trends and future perspectives. Comprehensive Reviews in Food Science and Food Safety. 2016;**15**(3):471-490. DOI: 10.1111/1541-4337.12200

[131] Liu Q, Wu JE, Lim ZY, Lai S, Lee N, Yang H. Metabolite profiling of *listeria innocua* for unravelling the inactivation mechanism of electrolysed water by nuclear magnetic resonance spectroscopy. International Journal of Food Microbiology. 2018;**271**:24-32. DOI: 10.1016/j.ijfoodmicro.2018.02.014

[132] Afari KG, Hung YC, King CH, Hu A. Reduction of *Escherichia coli* O157: H7 and *salmonella typhimurium* DT 104 on fresh produce using an automated washer with near neutral electrolyzed (NEO) water and ultrasound. Food Control. 2016;**63**:246-254. DOI: 10.1016/j.foodcont.2015.11.038

[133] Zhang C, Cao W, Hung YC, Li B. Application of electrolyzed oxidizing water in production of radish sprouts to reduce natural microbiota. Food Control. 2016;**2016**(67):177-182. DOI: 10.1016/j.foodcont.2016.02.045

[134] Turantaş F, Ersus-Bilek S, Sömek Ö, Alper KA. Decontamination effect of electrolyzed water washing on fruits and vegetables. Journal of Microbiology, Biotechnology and Food Sciences. 2018;**7**(4):337-342. DOI: 10.15414/jmbfs.2018.7.4.337-342

[135] Chen X, Xue SJ, Shi J, Kostrzynska M, Tang J, Guevremont E, et al. Red cabbage washing with acidic electrolyzed water: Effects on microbiological and physicochemical properties. Food Quality and Safety. 2018;**2**:229-237. DOI: 10.20944/preprints201808.0470.v1

[136] Ding T, Ge Z, Shi J, Xu YT, Jones CL, Liu DH. Impact of slightly acidic electrolyzed water (SAEW) and ultrasound on microbial loads and quality of fresh fruits. LWT-Food Science and Technology. 2015;**60**(2):1195-1199. DOI: 10.1016/j.lwt.2014.09.012

[137] Pinto L, Ippolito A, Baruzzi F. Control of spoiler *pseudomonas* spp. on fresh cut vegetables by neutral electrolyzed water. Food Microbiology. 2015;**50**:102-108. DOI: 10.1016/j.fm.2015.04.003

[138] Luo K, Oh DH. Inactivation kinetics of *listeria monocytogenes* and *salmonella enterica* serovar *typhimurium* on fresh-cut bell pepper treated with slightly acidic electrolyzed water combined with ultrasound and mild heat. Food Microbiology. 2016;**53**:165-171. DOI: 10.1016/j.fm.2015.09.014

[139] Zhang J, Yang H, Chan JZY. Development of portable flow-through electrochemical sanitizing unit to generate near neutral electrolyzed water. Journal of Food Science. 2018;**83**(3). DOI: 10.1111/1750-3841.14080

Chapter 9

Arousing Public Attention on Sea Level Rise in New Zealand through Art-Science Collaboration

Laura Donkers

Abstract

In New Zealand, climate scientists predict that climate change-induced sea level rise will have an earlier and greater impact on coastal communities than previously anticipated. In Auckland, the "City of Sails," Aucklanders' prize the opportunity to sail on the ocean and live near the beach. However, in 2019 Auckland Council released information that by 2060, a projected increase of 50 cm sea level rise would inundate the homes of 43,000 citizens. If citizens are to safeguard their lifestyles, they need to make effective decisions about how and where they choose to live. While artists are not often qualified to disseminate scientific knowledge, they are able to offer artistic comprehension through aesthetic intelligence, experientiality, and the creation of mental imagery. Building on this position, this chapter explores how an art-science exhibition, *Blue Radius,* deployed a range of sensorial, emotional, and scientific perspectives to imaginatively engage citizens with the phenomena of climate change-induced sea level rise and present relevant scientific information to assist citizens develop informed decision-making skills.

Keywords: art-science collaborations, climate change testimony, decision-making, imagination, aesthetic intelligence

1. Introduction

In this chapter, I hypothesise that artist-scientist collaborations can help to generate discrete engagement strategies that arouse public attention towards consciously and effectively engaging with climate and ecological science. The *Blue Radius* art exhibition (hereafter *Blue Radius*) was designed to raise public awareness of climate change-induced sea-level rise with the aim of exploring how art-science collaborations can help citizens to navigate, accept, and understand the future **to make better environment-influenced decisions.**

Human behaviours and activities have been changing the climate and ecology since industrialisation began. From disappearing ice to plastics in our food chain, in a little over 100 years we have speeded up the process of global warming by nearly doubling the volume of carbon in our atmosphere. This excess carbon is a major factor in causing average sea levels to rise at an alarming speed [1]. Also, as glaciers and ice sheets melt, the volume of the ocean is expanding as the water warms [2]. In addition to the problem of sea level rise, is the ongoing damage caused to marine ecology through activities like coastal sandmining for construction and marina construction on fragile ecological sites, which destroy the seabed environment [3, 4].

However, not all sea level rise is due to climate change. In countries that sit on tectonic plate boundaries, such as New Zealand, the coastline is also sinking due to "vertical land movement" (**Figure 1**) [5]. This is expected to accelerate the impact of climate change-induced sea level rise in high-population areas such as the Auckland region—currently 1,715,600 citizens with a population density of 347 people per km 2 (June 2021) [6]. Current safeguarding choices focus on coastline protection using sea walls and flood defences, adapting buildings and roads, or by retreating from "at-risk" areas [7]. However, environmental scientists, such as Professor Judy Lawrence and Dr. Paula Blackett, are calling for a "shift in mindset" where citizens support longer term solutions to protect future generations using relevant, accessible information to inform better decision-making [8].

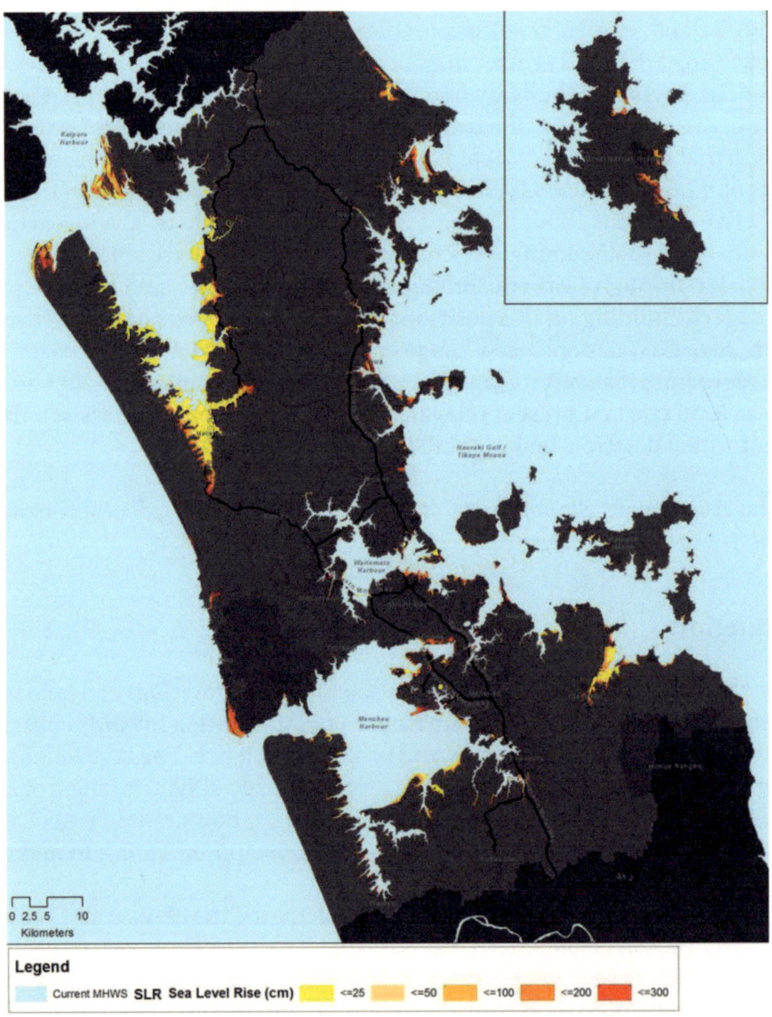

Figure 1.
Map of New Zealand showing areas impacted by sea level rise (Image Credit: Auckland Council).

It is beyond doubt that present modes of culture and lifestyles are contributing towards the climatic and ecological problems. But how society chooses to respond to these multiple ecological crises will determine its future [9]. In the most recent report by the International Panel on Climate Change (IPCC), experts stress that "if human behaviours do not change then existing climate trends will worsen" [10]. Therefore, if we are to safeguard the future, we must come to terms with our unsustainable relationship with nature. We need to learn about the consequences that have been unleashed and embrace ways to live more sustainably so that we do not endanger the opportunities for future generations, both human and nonhuman, to meet their basic needs. Sustainable development involves the creation of new things—such as environments, objects, and lifestyles—but also new visions and concepts that can support ethical, proactive, and optimistic future development [11].

This is where art can help "summon the future" by enabling us to perceive with new eyes, new minds, and new awareness [12]. But these possible futures must also include more holistic perspectives that show compassion and empathy for the creatures we live with. To this end, art can promote a change in attitude that involves "joy, ingenuity, self-respect, and a responsible attitude" [11], all of which can be highly effective in helping communities to see themselves as part of the nature/culture continuum [13].

Art helps to develop aesthetic intelligence, which is the human capacity to understand beauty; "where something deep within us is touched and fills us with delight" [14]. Aesthetic intelligence is a perceptual communication that arises through embodied and social experience: "a call-and-response between [self] and place" that occurs through experiences, memories, and making connections [15]. Ecological art highlights the interdependence between society and nature, commodities and matter. Its focus is on the interrelationships between the physical and the biological, cultural, political, and historical aspects of ecological systems. Art with an ecological focus seeks to reclaim, restore, or remediate natural environments. It informs the public about the environmental problems we face and aims to re-envision ecological relationships, proposing new possibilities for co-existence, sustainability, and healing [16, 17].

While artists are not often qualified to spread scientific knowledge, nor explain the environmental effects of decisions made in society they are able to offer artistic comprehension of life-based on aesthetic intelligence—through the experiential and observed, and the development of one's own mental images, public expression, and what can be visualised [11, 14]. When artists work with scientists they can create new awareness of the climate crisis and help citizens to perceive the future with new perspectives. "[T]he combination of science and artistic expression [can] bridge the emotional divide and penetrate the audience's consciousness with a glimpse of reality" [18]. "Critical and creative thinking [...] provide a unique and innovative way to engage with both the challenges and opportunities arising from climate change" [19].

2. Methodology

I used a collaborative art-science exhibition and engagement programme as a methodology to present a range of perspectives around the impacts that climate change induced sea level rise pose for the community. *Blue Radius* was presented at Depot Artspace, a community gallery in one of Auckland's many coastal suburbs for a four-week period from 3 to 28 September 2022. It comprised of exhibits by artists, activists, scientists, and organisations, and was the venue for several motivational talks by activists and scientists.

The exhibition aimed to present:

- connections between human activities and their impacts on the ocean and marine ecology
- adaptation options for citizens in response to sea level rise impacts
- the potential for creative engagement to help citizens understand more about the impacts of climate change induced sea level rise

2.1 Questions

This study was structured around answering the following research questions:

1. How can we develop more holistic perspectives to improve relationships with oceans and marine ecology?
2. How can we learn to engage with the loss of human control over our natural environment that climate change imposes?
3. What role can creative engagement play in helping us to explore these challenges?

2.2 Exhibition components

Blue Radius brings together the work of artists, activists, scientists, and organisations who reflected upon a range of perspectives as impetus for the public to consider their own positions and support to guide better decision-making about how they will continue to live in their places (**Figure 2**).

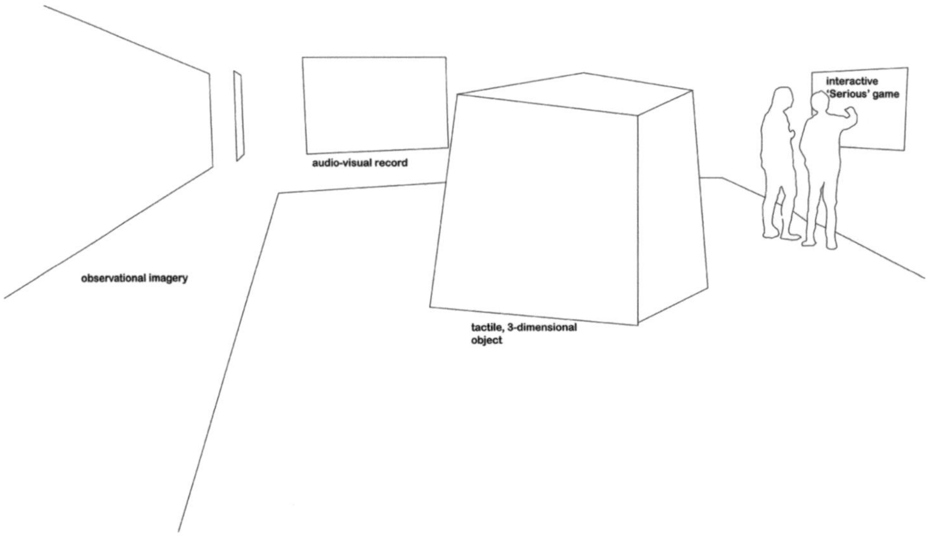

Figure 2.
Diagram identifying the various elements of the exhibition used to engage viewers (Image credit: Laura Donkers).

The exhibition introduced the following works:

- *Land Radius|2*—an audio-visual exchange between artists, scientists, community members, and activists who share testimony of their observations and frustrations about the ecological emergencies happening in the Haruaki Gulf Marine Park. Directed by Laura Donkers, this meditative film developed awareness of human interdependence with nature and individual perspectives affecting a range of environments and communities.

- *My Coastal Future*—a new "Serious Game" created especially for the exhibition by NIWA (New Zealand's National Institute of Water and Atmospheric Research), provided the player with the experience of decision-making about their coastal property as the sea level rises. The player could decide to build a seawall, move their house, or move elsewhere. The game introduced several models that include costings and worst case/least case scenarios to allow some elements of chance to enter the game and make it more of an experience for the player in the hope that key information would stick in their mind.

- *Coast Under Threat*—is a photographic essay by fine art photographer Stephen Perry that documents the deteriorating structures introduced by authorities and individuals as hard engineering to hold back nature. Piles of imported stones, concrete, and rusty steel are photographed far from the cliff face presenting evidence of their temporary nature. This misguided enthusiasm is repeated by current cliff-top dwellers who instal ever more elaborate constructions to hold back the elements. When will it stop?

- *Tuakana Teina*—takes center stage in the exhibition. A "Carbon Stack" created by Bianca Ranson and Te Aata Rangimarie Smith, forms a towering, tactile, three-dimensional object to stimulate pro-environmental consciousness. It promotes more sustainable ways of living through community food growing and taking an activist's approach highlighting the destruction of Kororā, Little Blue Penguin, habitats due to construction of a controversial 140-berth marina on Waiheke Island. This work involves 300 piles being rammed into the seabed, causing a section of active Kororā habitat to be destroyed.

- *Not Quite a Church | Inciting Public Gathering*—is the work of Nââwié Tutugoro, which references the raid on Camp Kororā, the activist's camp on Waiheke Island.

- *Ngā Aua Rere Kaharunga*—a woven sculpture by Atareta Black who weaves traditional Māori knowledge, genealogy, and traditional stories to convey relationships to the sea, land, and environment

3. Analysis

Blue Radius brought together art and science perspectives to drive home the immediacy of climate change—something so large, world-altering, and seemingly too distant to fathom—by looking to local manifestations of ecological collapse that we could be preventing. It was a collaboration between artists, scientists, organisations and businesses that presented unique sensorial, emotional and scientific perspectives to illustrate how human choice influences our ability to contemplate, respond to, and act on the ecological challenges we have created. The exhibition proposed that a more holistic perspective would enable people to confront how present modes of living diminish not only our lives but also the prospects of future generations.

The exhibition took place in Auckland, which is New Zealand's largest city. It is home to many harbours and marinas with moorings for more than 500,000 sailboats and yachts, giving rise to its nickname, the "City of Sails." With a coastline extending to 3200 km, living with the ocean is part of Aucklanders' identities, as an estimated 25% of households own at least one boat, and living with a beach view is highly sought after. However, sea level rise threatens this idyll. In 2019 newspapers were widely reporting new research released by Auckland Council that by 2060 a projected increase of 50 cm sea level rise, increased storm surges, high tides, and large waves would mean that the homes of more than 43,000 Aucklanders will be inundated and leave up to 30% of the city center at risk [20–22].

Considering this situation, *Blue Radius* sought to present climate change and human decision-making as the key issues prompting sea level rise and other ecological emergencies taking place along Auckland's coastlines. Spanning photography, sculpture, audio-visual installation, and video gaming, *Blue Radius* reflected on the function of art as a platform for critical engagement that juxtaposed the perspectives of artists and activists, scientists and organisations to examine the impact of sea level rise on the lives of citizens. It sought to change perceptions of climate and ecological crises by offering a range of spatial interventions for gallery visitors to navigate, using observational imagery, audio-visual testimony, an interactive "Serious" game, and the confronting spectacle of an eight-cubic meter Carbon Stack made of layered hay, coffee husks, and dried leaves, placed right in the middle of the gallery space (**Figure 3**).

At two-meters high, *Tuakana Teina* disrupted views across the gallery. Visitors were at times perplexed by its presence: "it's impossible to avoid, there it is—everywhere you look!", "a larger-than-life structure that gets in the way", "its spoiling my view of the other works", "it blocked my view of the speakers". The Carbon Stack concept belongs to Richard Wallis, founder of The Carbon Cycle Company, and inventor of the Carbon Composter who helps individuals, companies, and communities to reduce CO_2 emissions by composting local food waste. Having a store of carbon

Figure 3.
Blue Radius exhibition view (Image Credit: Laura Donkers).

beside your community compost heap ensures that local food waste can be properly composted down to a beneficial growing medium.

Wallis created a composting system with Bianca Ranson at the community garden at her local *marae* in Waiheke Island. Ranson chose to use the idea of the Carbon Stack as a metaphor for healing environmental and societal ills, identifying that "Climate justice starts with those most vulnerable in our community including our taonga (treasure) species. Extraction, pollution, and governance have left our moana (ocean) in a biodiversity crisis, facing ecological collapse. The mauri (life force) of our moana is under threat."

Her "elephant-sized" Carbon Stack is decorated with a shredded 1000-page court injunction and a Pohutukawa tree that was ripped out by the private developers who were building the unwelcome luxury 140-berth marina. She, along with 31 other members of her community, were each sanctioned with a trespassing injunction and given a $750,000 fine for swimming in their own bay in protest at the construction work that destroyed the habitats of the nesting Kororā (native Little Blue Penguin). The concept of her work, *Tuakana Teina,* highlights the Māori belief that all valued species such as the Kororā "come before us in our whakapapa (genealogy). Our relationship to them is tapu (sacred) and we have an inherent responsibility to protect them." The Carbon Stack contributes to the composting of all food waste on Waiheke Island. When these artefacts become nutrient-rich soil and food, they will feed the hungry stomachs of those making this urgent stand at a time of climate crisis and biodiversity loss (**Figure 4**).

NIWA's *My Coastal Futures Game* afforded a very different experience to visitors. Its accessible format invited participation, which was for all ages from 10 years upwards. The game took you through a decision-making process in around 5 minutes. Its purpose was to help players understand some of the science of climate change and the things we can do to combat it both individually and, with their families, schools and communities (**Figure 3**). The idea was to bring the impact of changing climate, rising seas, and worsening storms that damage the coastline into the hands of the players to make decisions about their coastal property as the sea level rises. All is not lost, there are ways you can adapt. You can build a seawall, move your house back on the section, or move elsewhere. What you choose is up to you but watch out—things can change quickly! Spend your money wisely to determine your coastal future.

NIWA, the National Institute of Water and Atmospheric Research, is New Zealand's Crown Research Institute for climate, freshwater and marine science. They provide decision makers with the science-based information they need to make informed choices in a changing world.

3.1 Effective decision-making around sea level rise impacts

Blue Radius presented a local exploration of the ecological emergencies taking place along Auckland's coastlines and how human activity was inducing climate breakdown. It looked at the ways humans damage local marine ecology through ongoing extractive practices by mining coastal sand for construction that destroys seabed environments, while permits are granted for marina construction on fragile ecological sites. The exhibition proposed that a more holistic perspective would confront how our present modes of living diminish not only our lives but also the prospects of future generations. Building on our human capacities to innovate is an important way to motivate action [12] but our innovative capacities must include compassion

Figure 4.
Family playing my coastal futures game (Image credit: Laura Donkers).

and empathy for the non-human entities we live with and rely on for our continued prosperity.

Spanning photography, sculpture, audio-visual installation, and video gaming, *Blue Radius* reflected on the function of art as a platform for critical engagement that juxtaposed the perspectives of artists and activists, scientists and organisations to examine the impact of sea level rise on the lives of citizens. It sought to change perceptions of climate and ecological crises by offering a range of spatial interventions for gallery visitors to navigate, through observational imagery, audio-visual testimony, a towering, tactile, three-dimensional object, and a relevant, interactive "Serious" game. The exhibition explored how individual and collective voices can help to illuminate structures of policy and decision-making and give voice to the non-human entities that exist and suffer because of destructive human activity. This multi-disciplinary approach supported connection-building across human and non-human communities to imagine, engage, and inform the public.

4. Conclusion

Exhibitors presented visualisations that imaginatively explored aspects of the climate and other emergencies. Video, photography, sculpture, and an online, interactive Serious Game presented the impacts of decision-making that result in eroded favourite beaches, ruined shellfish colonies, and destroyed habitats such as that of the Kororā, the Little Blue Penguin. The interrelated artworks enabled visitors to reflect on and process different experiences of loss that are expected to be experienced in the future: the loss of favourite beaches, loss of homes and lifestyles, loss of species, loss of community.

Blue Radius created a platform for an innovative and experimental exchange of views where culture, community and science could meet in a non-hierarchical dialogue. The novelty of this approach lay in the explorative nature of where the process could lead. The goal was to develop imaginative strategies that would help to improve understanding, increase levels of engagement and help to share experiences of climate change induced sea level rise and other issues affecting marine ecology, in the hope that those present might be willing to consider the impact our behaviours and attitudes are having on nature.

Thus, *Blue Radius* contributed to knowledge exchange across art, sustainability education, ecology, and environmental disciplines. It connected with a broad cross section of citizens who might be struggling to focus their efforts in a world of biodiversity loss and climatic change, such as children and young people, teachers, local communities, indigenous communities, artists, politicians, local authorities, business owners, tourists, new immigrants, and community organisations. Further to these associations, this exhibition provided opportunities to extend the reach of artistic production by building awareness, decision-making, education, and empowerment to encourage local communities to undertake pro-environmental action.

To conclude, *Blue Radius* provided a public platform for a diverse group of artists, activists, scientists, and organisations whose work imaginatively explored the phenomena of climate change induced sea level rise. The exhibition introduced creative, ecological, and science-based methods to develop awareness of human interdependence with nature. It stimulated pro-environmental consciousness by promoting more sustainable ways of living. It gave voice to imaginative and moral states of mind towards nature. It developed new collaborations that helped to unlock new forms of inquiry. It explored how individual and collective voices can help to illuminate structures of policy and decision-making and give voice to the non-human entities that exist and suffer because of destructive human activity. This multi-disciplinary approach supported connection-building across human and non-human communities to engage and inform the public.

Author details

Laura Donkers
Independent Artist-Curator-Researcher, Auckland, New Zealand

*Address all correspondence to: laura.donkers.art@googlemail.com

References

[1] Sweet WV, Hamlington BD, Kopp RE, Weaver CP, Barnard PL, Bekaert D, et al. Global and Regional Sea Level Rise Scenarios for the United States: Updated Mean Projections and Extreme Water Level Probabilities Along U.S. Coastlines. NOAA Technical Report NOS 01. Silver Spring, MD: National Oceanic and Atmospheric Administration, National Ocean Service; 2022. p. 111 Available from: https://oceanservice.noaa.gov/hazards/sealevelrise/sealevelrise-tech-report.html

[2] Lindsey R. Climate Change: Global Sea Level. Understanding Climate, NOAAClimate.gov. 14 August 2020. Available from: https://www.climate.gov/news-features/understanding-climate/climate-change-global-sea-level [Accessed July 31, 2022]

[3] Muru-Lanning C. The occupation at Pūtiki Bay, Waiheke – explained. The Spinoff. 2021 June 23 Available from: https://thespinoff.co.nz/atea/23-06-2021/the-occupation-at-putiki-bay-waiheke-explained [Accessed August 2, 2022]

[4] Vidyadharan A. Save our sands: Hundreds gather to oppose sand mining at Pakiri. New Zealand Herald. 2022, July 10 Available from: https://www.nzherald.co.nz/northern-advocate/news/save-our-sands-hundreds-gather-to-oppose-sand-mining-at-pakiri/5HHJSBW2S2LJOEMDL4NRWEDGLQ/

[5] Meduna V. A map of the future. Science & Environment. 2022 Available from: https://www.nzgeo.com/stories/a-map-of-the-future/

[6] Wikipedia Auckland Region. 2022. Available from: https://en.wikipedia.org/wiki/Auckland_Region#:~:text=Auckland%20Region%20covers%204%2C941.13%20km%202%20%281%2C907.78%20sq,population%20density%20of%20347%20people%20per%20km%202. [Accessed July 31, 2022]

[7] Small Z. 'There's an opportunity now': NZ's three 'big choices' to safeguard against sea level rise. News Hub, NZ. 2022, May 03 Available from: https://www.msn.com/en-nz/news/national/there-s-an-opportunity-now-nz-s-three-big-choices-to-safeguard-against-sea-level-rise/ar-AAWRdCc?ocid=msedgntp&cvid=c835ee79fe954349b9cda1fb47f4f801

[8] Lawrence J, Stephens S, Blackett P, Priestley R. Climate services transformed: Decision-making practice for the coast in a changing climate. Frontiers in Marine Science. 2021. Available from: https://www.frontiersin.org/articles/10.3389/fmars.2021.703902/full

[9] Figueres C, Rivett-Carnac T. The Future We Choose: Surviving the Climate Crisis. London: Manilla Press; 2020

[10] IPCC (2022) Summary for policymakers. In: *Climate Change 2022: Mitigation of Climate Change. Contribution of Working Group III to the Sixth Assessment Report of the Intergovernmental Panel on Climate Change* [P.R. Shukla, J. Skea, R. Slade, A. Al Khourdajie, R. van Diemen, D. McCollum, M. Pathak, S. Some, P. Vyas, R. Fradera, M. Belkacemi, A. Hasija, G. Lisboa, S. Luz, J. Malley, (eds.)]. Cambridge University Press, Cambridge, UK and New York, NY, USA. doi: 10.1017/9781009157926.001 Available from: https://www.ipcc.ch/report/ar6/wg3/downloads/report/IPCC_AR6_WGIII_SPM.pdf

[11] Mantere M-H. Ecology, environmental education and art

teaching. In: Piironen L, editor. Power of Images. Helsinki, Finland: INSEA Research Congress, Association of Art Teachers; 1992. pp. 17-26 Available from: http://natureartеducation.org/Articles/Ecology,%20Environmental%20Education%20and%20Art%20Teaching.pdf

[12] Eagleman D, Brandt A. The Runaway Species: How Human Creativity Remakes the World. Edinburgh: Canongate Books; 2017

[13] Demos TJ. Decolonizing Nature: Contemporary Art and the Politics of Ecology. New York: Sternberg Press; 2016

[14] Mucha R. Aesthetic intelligence cultivating an artistic mindset. *Organization Development Journal*; Chesterland. 2008;**26**(3):41-54. Available from: https://www.proquest.com/docview/198043939?pq-origsite=gscholar&fromopenview=true [Accessed July 31, 2022]

[15] Burrill RR. Art as ecology: A mutual nod. The Ecological Citizen. 2022;**5**(2):178-185 Available from: https://www.ecologicalcitizen.net/pdfs/epub-067.pdf?fbclid=IwAR2K3uEVpD_yOli1STQXMrYFJZ_OxFrZTBC1MVV3LmCkfUKcxTX0wProNaM

[16] Kagan SJ. The practice of ecological art – [plastic]. 2014. Available from: http://art-science.univ-paris1.fr/plastik/document.php?id=866

[17] Geffen A, Rosenthal A, Freemantle C, Rahmani A, editors. Ecoart in Action Activities, Case Studies, and Provocations for Classrooms and Communities. New York: New Village Press; 2022

[18] Peters CH. Art and science collaboration produces work-in-progress on climate change. Princeton University (Web Blog). 6 April 2010. Available from: https://www.princeton.edu/news/2010/04/06/art-and-science-collaboration-produces-work-progress-climate-change

[19] De Freitas W, Abrahams G. Climate science is looking to art to create change. *The Conversation*. 2015, May 6 Available from: https://theconversation.com/climate-science-is-looking-to-art-to-create-change-41185

[20] Morton J. Auckland's rising seas: Insurance warning as 43,000 at risk. NZ Herald. 24 Mar 2019. Available from: https://www.nzherald.co.nz/nz/aucklands-rising-seas-insurance-warning-as-43000-at-risk/F4JO64DI2MQOXZK2G6GAJ2ISFE/

[21] Dickens A. Auckland's rising seas: Insurance warning as 43,000 at risk. Newstalk ZB. 25 Mar 2019. Available from: https://www.newstalkzb.co.nz/on-air/andrew-dickens-monday-afternoons/audio/talkback-callers-react-to-report-on-sea-levels-rising/

[22] Niall T. Climate change: Rising sea level could hit 43,000 Aucklanders' homes Stuff. 22 Mar 2019. Available from: https://www.stuff.co.nz/environment/climate-news/111472176/climate-change-rising-sea-level-could-hit-43000-aucklanders-homes

Section 4

Adaptation

Chapter 10

Adaptation Strategies for Climate Variability in the High Rainfall Zone of India, Assam

Pallab Kumar Sarma, Nikhilesh Baruah, Rupam Borah, Rupshree Borah, Arunjyoti Sonowal, Rekhashree Kalita, Prasanta Neog, Prabal Saikia and Nipen Gogoi

Abstract

The NICRA project is being implemented in two villages *viz.*, Chamua (since 2010–2011) and Ganakdalani (since 2012–2013 till 2016–2017), which are situated in the west of Lakhimpur district of North Bank Plains Zone of Assam. Chamua village is situated in Kherajkhat Mauza (Taluka), which is 45 km away from North Lakhimpur, the headquarter of district Lakhimpur. On the other hand, Ganakdoloni is situated at Dhalpur Mauza, situated 60 km away from North Lakhimpur and 15 km away from the local township Narayanpur. During 2017–2018 four villages *viz., Jakaipelua, Borbali, Borkhet, and Nogaya* were adopted under the project. Analysis of long-term rainfall data confirmed the significant decreasing trend of annual as well as monsoonal rainfall in both the Brahmaputra and Barak basins of Assam, India. Variability of rainfall has been increasing in terms of the increased frequency of high-intensity rains and the reduced number of rainy days, leading to localized flash floods and the occurrence of multiple dry spells. Mean season-wise rainfall 2011–2021 indicates long dry periods during the winter season, leading to prolonged dry spells affecting crop growth. About 69% of total rainfall (average annual rainfall of Assam is 2000 mm) is received during the monsoon season, resulting in flash floods leading to crop damage. Out of 12 years of investigation, 10 years are deficit years, resulting in crop stress both during the monsoon and post-monsoon period. Preparation and implementation of real-time crop contingencies are important in responding to weather aberrations in different strategies like preparedness, real-time response, etc. Identification of various adaptation strategies, including climate-resilient crops and cultivars, rainwater harvesting and recycling, efficient energy management through farm mechanization, dissemination of weather information, and weather-based agro-advisories to farmers in a real-time basis, is important adaptation technologies for building climate-resilient agriculture. The study showed that adaption of climate-resilient crop and cropping system and use of harvested rainwater resulted in a 12 to 30% increase in yield observed by the cultivation of high-yielding rice varieties (HYVs) (*Ranjit, Gitesh, Mahsuri*, etc.) when sown in time (before 15th June) over late sowing conditions (after 20th June). In the case of early season drought, replacement of long duration traditional varieties with short duration HYV and life-saving irrigation using harvested rainwater increased yield by about 59% (short duration var. *Dishang*) over non-irrigated fields. In case of mid-season and terminal drought, application of an additional dose of 22 kg ha^{-1} MOP at maximum tillering to grain growth period an increase in yield of about 33% (*Ranjit*), 32% (*Gitesh*), 64% (*Shraboni*), and 57.5% (*Mulagabharu*) has been observed over farmers' practice. In highly flood-affected areas under lowland situations replacement of submergence tolerant varieties (*Jalashree* and *Jalkuwari*) with traditional deepwater rice varieties resulted in reduced crop loss due to the genetic trait of the deepwater rice, which can withstand water logging for a long period. With an increase in the level of mechanization through the use of machinery available in the custom hiring center the human and animal hour requirement for paddy cultivation was reduced from 795 to 350 $hrha^{-1}$ and 353 to 23 $hrha^{-1}$, respectively. Alternate land use in terms of low-cost poly house, vermicompost production, and mushroom cultivation also resulted in nutritional security and generation of higher income for the farmers

Keywords: climate change, ecosystem, extreme weather, vulnerability, National Innovations on Climate Resilient Agriculture (NICRA), North Bank Plains Zone (NBPZ) of Assam

1. Introduction

Climate change has become an important area of concern for India to ensure food and nutritional security for the growing population. Climate change has been one of the most talked about subjects by both scientists and common people today. The major concern has been the question of food security soon under the changed climatic situation. Climate is changing and this change is now for real. It has come as a daunting challenge to agriculture and agriculturists. The impacts of climate change are global, but countries like India are more vulnerable because of the high population depending on agriculture. The Government of India has accorded high priority to research and development to cope with climate change in the agriculture sector. The Prime Minister's National Action Plan on climate change has identified agriculture as one of the eight national missions. Therefore, the time has come to shift our focus from assessing the impact on agriculture to the management-based solution to cope with food production sustainability. Many researchers are underway to evolve climate-friendly, climate-smart, and climate-neutral agricultural technology for the benefit of society.

The impact of climate change has been increasingly visible in terms of change in temperature, precipitation, retreating ice caps and glaciers, sea level rise, atmospheric circulation pattern, and ecosystems. Climate change is not simply the increasing temperature but is indeed responsible for the increasing frequency of extreme weather events, such as heat waves, cold waves, droughts, and floods. In India, the increased frequency of pronounced heat waves, cold waves, droughts, and floods has already been realized in the last few decades. Analysis of long-term rainfall data confirmed the significant decreasing trend of annual as well as monsoonal rainfall in both the Brahmaputra and Barak basins of Assam, India [1]. Variability of rainfall has been increasing in terms of the increased frequency of high-intensity rains and the reduced number of rainy days, leading to localized flash floods and the occurrence of multiple dry spells [2]. The chapter will deal with strategies to adapt to climate variability from farmers' perspectives based on the experiences of the All-India Network Project on

Class	All classes	
	Number	%age
Marginal (<1 ha)	1,669,252	62.2
Small (1.0–2.0 ha)	561,078	20.9
Semi medium (2.0–4.0 ha)	351,245	13.1
Medium (4.0–10.0 ha)	96,418	3.5
Large(> 10.0 ha)	5004	0.2

Table 1.
Number of land holdings according to size classes of Assam [3].

National Innovations on Climate Resilient Agriculture (NICRA) being implemented in the North Bank Plains Zone of Assam (NBPZ), India since 2011.

1.1 Physiography of Assam

Assam is one of the states of northeast India situated between 24°and 28°16′N latitude and 89°4′and 96°E longitude. It is surrounded by Bhutan and Arunachal Pradesh in the north, Arunachal, Nagaland, and Manipur in the east, Mizoram, Tripura, and Meghalaya in the south, and Bangladesh, Meghalaya, and West Bengal in the west. Assam has a geographical area of 78,523 square km comprising fertile land and water resources. Despite these resources, Assam is lagging in the agriculture sector. The farming community mostly belongs to small (20.9%) and marginal (62.2%), with large farmers being very low (0.19%). The average farm size is decreasing over the years (1.11 ha) (**Table 1**) [3].

The physiography of the state can be divided into three distinct units: the plain, the plateau, and the hills. Two primary units for agricultural development in Assam are the Brahmaputra and the Barak valley, accounting for 80.8% of the total geographical area. The state of Assam has been divided into six agro-climatic zones based on physiographic variation, climate, soil, farming systems, crops and cropping systems, and hydrology under the National Agricultural Research Project (NARP) (**Table 2**).

1.2 Soil and climate of Assam

The major portion of the soils of Assam belongs to Inceptisols (49.3%) followed by Entisols (32.3%), Alfisols (12.3%), and Ultisols (6.1%). The texture of the soils varies from sandy loam to clay loam depending on the agroclimatic conditions and physiographic units. The soil of Assam is acidic having a pH range from 4.2 to 5.8. The average status of available nitrogen, phosphorus, and potassium status in the soil is low to medium, with a slight variation in the content [5, 6]. The soil of the state as a whole is deficient in boron and marginal in the case of zinc, iron, and aluminum toxicity are also observed sporadically.

The climate of Assam is humid subtropical in nature with warm humid summer and cool dry winter. The rainfall in Assam is in general high, but its distribution over time and space is not uniform and even. The monsoon period is characterized by high-intensity rainfall, while the winter (December to February) is virtually dry. Flood is a regular feature affecting the *Kharif* rice (winter paddy) production in about 7.75 Lakh ha of agricultural land annually. The mean annual maximum temperature

Sl. No.	Agro-Climatic Zones	Number of Districts	Net cropped area(ha)	Area sowed more than once	Cropping Intensity (%)
1	Lower Brahmaputra Valley Zone	10	9,29,757	4,69,422	150
2	North Bank Plain Zone	5	5,36,598	3,36,323	163
3	Central Barak Valley Zone	2	3,27,637	85,615	126
4	Upper Brahmaputra Valley Zone	5	6,20,320	1,94,629	131
5	Barak Valley Zones	3	2,41,715	1,01,989	142
6	Hill Zones	2	1,54,570	1,00,824	165
	Assam	**27**	**28,10,597**	**12,88,865**	**146**

Table 2.
Agro-climatic zones of Assam [4].

lies between 23.6°C and 31.7°C, while the mean minimum temperature varies from 10°C to 25.2°C. The interception of bright sunshine is only 36 to 38% of the astronomically possible sunshine hours from June to August due to continuous overcast sky. Though sunshine hours are the longest from November to February (70 to 74%) radiation is not up to the desired level due to foggy weather.

1.3 Flood

A flood is a natural calamity where the state is exposed to vagaries of climate to an extreme scale, endangering the life and property of the farmers. The state is home to a mega network of rivers comprising 48 major and 128 small rivers, which are responsible for the annual floods, resulting in an average annual loss of Rs.200 crores for the state. Horticulture crops occupy 7.57 and 15.25% of total land and total cropped area, respectively.

1.4 Land use pattern

Assam has a total geographical area of 78,523 square km, of which 24.91% is under forest. The net cropped area constitutes 35.1% of total land area and 70.66% of total cultivated area. About 58.5% area is mono-cropped, especially with winter rice, with the cropping intensity being 146% (RKVY-Assam, 2022). About 186 lakh hectares and 1.51 lakh hectares are remaining fallow and waste. The state has a population of 31.17 million and the population of the state is expected to swell to 35.6 million by 2025, putting tremendous pressure on land and water resources.

1.5 Irrigation status in Assam

Irrigation is very critical in terms of nullifying the ill effect of weather vagaries, getting a response to input and other management practices, and exploiting the genetic potential of modern varieties. Irrigation has gained much importance given weather extremities line flood and drought-like situations frequently occurring as a result of climate change. At present only 18.5% of the net cultivated area of the state is under irrigation.

1.6 Present status of rainfall and temperature variability

1.6.1 Temperature

According to the fourth assessment report of the Intergovernmental Panel on Climate Change (2021), the increase in global surface temperature from 1850 to 1900 is assessed to be 1.09 [0.95 to 1.20] °C [7]. In the northeast region of India, the annual mean temperature is reported to be rising at a rate of 0.04°C per decade [8]. Assam also exhibits warming trends throughout with minor spatial variations. Analysis of temperature data during 1961–2010 reveals a warming trend in both mean maximum and minimum temperature. The magnitude of minimum temperature was higher compared to those of maximum temperature. Similar trends were also observed by other researchers [9].

1.6.2 Rainfall scenario of Assam

Assam experiences an average annual rainfall of 1980 mm, while during pre-monsoon (March–May), monsoon (June–September), post-monsoon (October–November), and winter period (December–February) average rainfall received is 486.5 mm (24.57%), 1279 mm (64.64%), 150.3 mm (7.59%), and 63.3 mm (3.20%), respectively (**Table 3**).

Deka and Mahanta reported a declining tendency of annual rainfall at a rate of 7.9 mm per 100 years in the NE region of India [10]. The summer monsoon rainfall in the region is decreasing (−48.0 mm/100 years), while rainfall during the post-monsoon season exhibited an increasing trend (+36.3 mm/100 years). The contribution of June, August, and September rainfall to annual rainfall exhibits a decreasing trend, while the contribution of July and October rainfall marks a rising trend [11]. On an annual basis, a long-term decreasing trend of rainfall has been observed in both the Brahmaputra basin and the Barak basin of Assam. Decreasing tendency (39.1 mm/decade) in the Barak valley, while in the Brahmaputra basin the annual decreasing trend was 9.6 mm per decade also reported [11]. About 72 million hectares of net sown area in India is rainfed, which is practiced in diverse climates and agro-ecologies contributing to about 40% of

Rainfall	Normal rainfall	Normal rainy days (No.)
Pre-monsoon (March–May)	486.5	32
Monsoon (June–September)	1279.0	56
Post-monsoon (Oct–Nov)	150.3	9
Dry periods (Dec–Feb)	63.3	8
Total	**1979.1**	**105**

Table 3.
Mean season-wise annual rainfall and rainy days in NBPZ of Assam [5].

the country's food basket. The agricultural production, productivity, and stability in rainfed areas are more vulnerable to climate variability, particularly during the kharif season, due to high dependency on the southwest monsoon. Monsoon failures result in a drought that has serious implications for small and marginal farmers. The annual average rainfall and the seasonal monthly distribution of rainfall for two districts of Assam (the study area and the adjacent district of the domain area) situated in the NBPZ of Assam had been shown in **Figures 1–4**, respectively.

1.7 Crops and cropping systems

See **Tables 4** and 5.

1.8 Climate change in India and Assam

Climate change and its variability are emerging as major challenges facing Indian agriculture. The high inter and intra-seasonal variability in rainfall distribution, rainfall events, and extreme temperature are causing crop damages and losses to the farmers [12]. In many parts of India, the frequency of occurrence of cold nights declined, while the frequency of occurrence of warm nights and warm days significantly

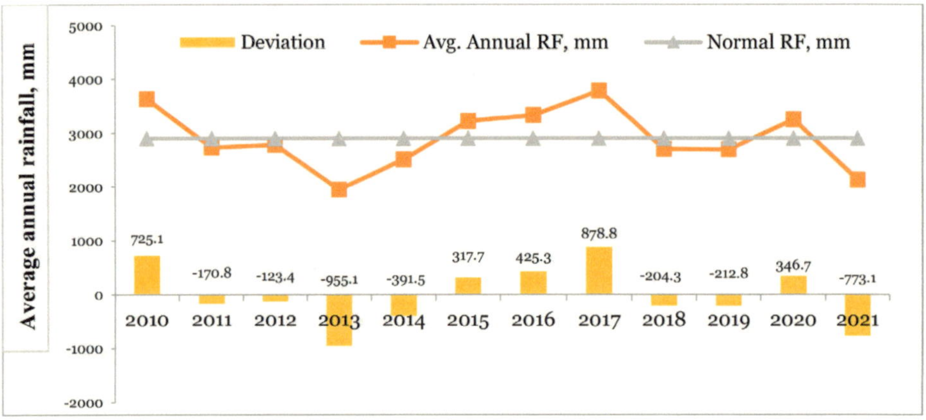

Figure 1.
Average annual rainfall, normal rainfall, and the deviation for Lakhimpur district of Assam (2010–2021).

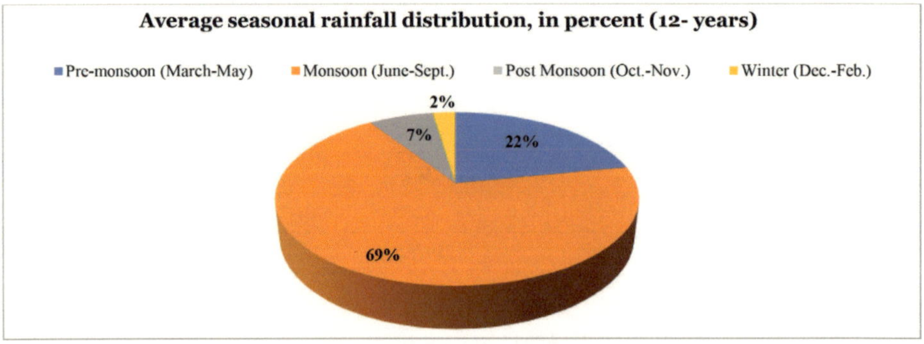

Figure 2.
Average seasonal rainfall for the Lakhimpur district of Assam (2010–2021).

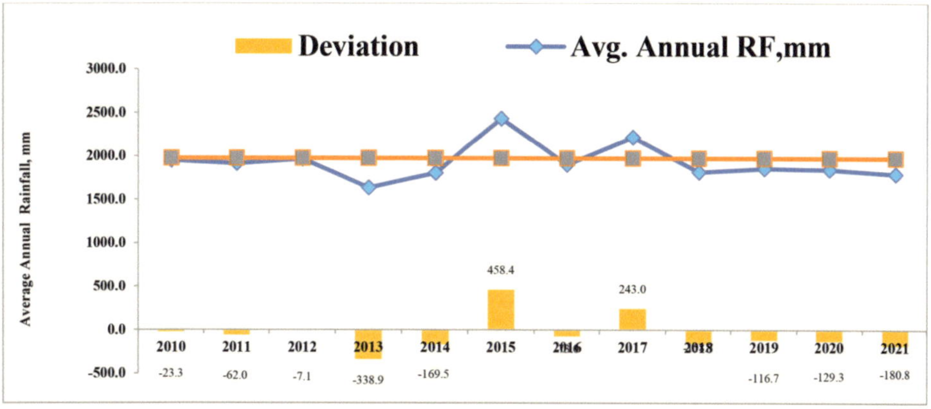

Figure 3.
Average annual rainfall, normal rainfall, and the deviation for Biswanath district of Assam (2010–2021).

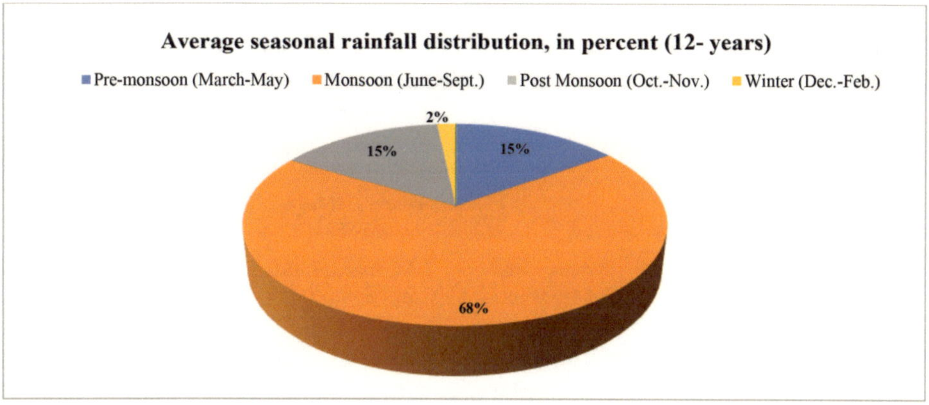

Figure 4.
Average seasonal rainfall for Biswanath district of Assam (2010–2021).

increased. Overall climate change pattern has already begun affecting the Indian agriculture sector adversely by enhancing abiotic and biotic stress to the crops and livestock. Various future climate models indicate a consistent warming trend over India, in short, mid as well as long-term scenarios [13]. Overall, the temperature rise is likely to be much higher during the *rabi* (October/November to January/February) than the *Kharif* (June/July to October/November) season. In addition, droughts and floods as well as cold and heat waves are likely to increase due to increased variability in temperature and may cause crop losses up to 30% by 2080. Assam also exhibits warming trends throughout with minor spatial variations. Analysis of temperature data from 1961 to 2010 revealed a warming trend in mean maximum temperature over Dibrugarh (0.120°C) per year, Jorhat (0.006°C per year), Nagaon (0.007°C/per year, Diphu (0.001°C per year and Guwahati 0.016°C. Tezpur, however, showed a decreasing trend of mean maximum temperature at the rate of 0.007°C per year. The magnitudes of trends of minimum temperature were higher compared to maximum temperature. The long-term (1901 to 2010) mean annual rainfall of the Brahmaputra basin is 2293 mm with a standard deviation of 225 mm. Barak valley basin receives 3204 mm rainfall with a stand's deviation of 420 mm. On an annual basis, a long-term decreasing trend of rainfall has been observed in both the basins [11].

Crop	Season	Sowing time	Variety	Duration (days)	Average yield (qha^{-1})
Direct seeded early *Ahu*	Autumn	Mid-February	*Luit, Kapilee,* IR 50, IR 36, *Dishang*	110–120	20–25
Transplanted early *Ahu*	Autumn	Mid-February	*Luit, Kapilee, Govind,* IR 36, *Rasi,* IR 50, *Jaya*	125–135	25–35
Normal *Ahu* direct seeded	Autumn	March/April	*Rasi,* IR 36, *Govind, Banglami, Rangadoria, Maibee*	105–120	25–30
Normal *Ahu* transplanted	Autumn	March/April sowing in nursery	*Lachit, Chilarai, Gopinath, Luit*	125–140	30–35
Sali Rice	*Kharif*	May/June in nursery	*Ranjit, Bahadur, Maniram, Mahsuri, KetekiJoha*	130–150	45–60
Boro	*Rabi*	November/ December sowing in nursery	*Joymoti, Kanaklata, Dinanath, Swarnabh*	160–170	50–65
Bao	Autumn	March/April	*Panindra, Padmanath, Maguri, Panchanan, Padumoni*	200–240	25–35
Maize	*Kharif*	March–May	*Ganga, Vivek maize* hybrid 47, *Vivek maize* hybrid 53, *Novjot,* Bio-9544	110–130	45–50
Greengram	*Kharif*	Mid-August to Mid-September	T-44, *Pratap,* K-851, ML-56, ML-131, *Kopergaon*	60–70	10–12
Blackgram	*Kharif*	Mid-August to Mid-September	T-9,T-27, Pant U-19, T-122, SB-121	80–90	10–12
Sesamum	*Kharif*	July-1st fortnight of August	*Madhavi, Gouri, Vinayak,* ST-1683	70–85	8–9
Toria	*Rabi*	Mid-October to Mid-November	TS-36, TS-38, TS-46, TS-67	90–95	10–12
Jute	Autumn	Mid-March to May	*Sonali, Shyamali, Bohagi, Navin*	110–120	22–26*
Pea	*Rabi*	Mid-October	*Rachna,* Pant 14	120–125	10–12
Sugarcane	Autumn	February /March	*Daria, Kalang, Barak, Dhansiri, Luhit*	280–300	700–800**
Potato	*Rabi*	October/ November	*Kufri Pokhraj, Kufri Jyoti, Kufri Megha*	80–110	85–100

*Fiber yield.
**Cane yield.

Table 4.
Major rainfed crops of Assam [5].

As per the findings of the State Action Plan for Climate Change, the annual mean temperature has increased by 0.59°C and the annual rainfall has decreased by −2.96/ mm per year over the last six decades (1960–2000). Climate change projections for Assam indicate that the mean average temperature is likely to rise by +1.7–2.2°C by

DOI: http://dx.doi.org/10.5772/intechopen.107045

Kharif-rainfed crops						*Rabi*-rainfed crops						
Rice	**Maize**	**Blackgram Greengram**	**Jute**	**Sugarcane**	**Sesame**	**Toria**	**Potato**	**Lentil**	**Pea**	**Rajmah**	**Sugarcane**	*Boro* **Rice**
Sali: June/July	Feb–April	Mid-Aug to Mid-Sept (Kharif)	March–May	Feb–March	July 1st fortnight of August	Mid-Oct to Mid-Nov	Mid-Oct to Mid-Nov.	Mid-Oct to Mid-Nov.	Mid Oct	Mid-Oct to Nov.	Oct	Nov-Dec
Ahu: March/April	—	Mid-Feb to March (Summer)	—	—	—	—	—	—		—	—	—
Bao rice: March/April			—	—	—	—	—	—	—	—	—	—

Table 5.
Normal sowing window of the rainfed crops.

mid-century between 1971 and 2000. There is likely to be an increase in extreme rainfall events by +5 to 38%, including floods. There is a chance of an increase in droughts weeks are going to rise as well, by more than 75% concerning the baseline (1971–2000) in the southern districts of Assam. As regards floods, projections increase the rise in events by more than 25% [14].

1.9 Impact of climate change on agriculture of Assam

The impact of climate change on agriculture is both direct and indirect. The direct impact of climate change would be small on rainy season crops but the crops will become vulnerable due to increased incidence of weather extremes, such as changes in rainy days, rainfall intensity, duration and frequency of drought and flood, diurnal asymmetry of temperature, change in humidity, and pest incidence and virulence. Winter crop production may become more vulnerable and the climate impact on cereals will vary widely in the rainy season as well as the winter season. Rainfed agriculture is likely to be more vulnerable because of the high dependency on monsoon and aberrant behavior of the south-west (SW) monsoon [15].

1.10 Farm mechanization

Agricultural productivity goes hand in hand with the mechanization of different agricultural operations, which aims at achieving timeliness of operations, efficient use of inputs *viz.,* seed, fertilizer, and chemicals, *etc.*, improvement in quality of produce, safety and comfort of farmers, reduction in the cost of produce and drudgery of farmers [16]. Mechanization would increase land and labor productivity and reduce the drudgery of humans and animals. In the changing climatic environment, the frequency of destructive weather aberrations (such as heat waves, heavy rainfall events, drought, and floods) affects the agriculture sector more vulnerable. In such situations, timely completion of farm operations is very important, which could be accomplished through the use of improved implements and machines. Thus, mechanization is the key to building climate-resilient agriculture in the country [17, 18].

Assam has the largest cultivable plain land in northeast India but power consumption for mechanization is 0.75 $kWha^{-1}$, which is still below the national average of 1.5 $kWha^{-1}$. For sustainable food grain production and drudgery reduction, the mechanization of agriculture is mandatory to an optimum level. Three main sources of farm power being utilized for these tools, machines, and equipment are manual (human), draft animal, and mechanical power. In many developing countries, up to 80% of farm power is provided by animal sources of power (**Table 6**) [18].

1.11 Real-time contingency planning–concept

Real-time contingency planning is considered as "any contingency measure either technology related (land, soil, water, and crop) or institutional and policy-based, which is implemented based on real-time weather pattern (including extreme events) in any crop growing season" [19]. Real-time contingency planning (RTCP) was conceptualized to minimize crop production and productivity losses and to improve the efficiency of the rainfed production system. The major objectives of RTCP are:

Operation	Implement/machinery	
	Bullock drawn/manual	**Tractor drawn**
Land preparation	Indigenous plow, mold board plow	Mold board plow, 2 bottom mold board plow, disc plow, heavy-duty disc harrow
Tillage	Indigenous plow, ladder	Harrows, cultivators, clod crushers, levelers, rotary tillers, rotavators, puddlers
Sowing	Broadcasting, transplanting, manual rice transplanter	Rice transplanter, self-propelled paddy transplanter
Residue incorporation		Disc plow/ harrow
Interculture	*Bindha*/weeder, hand hoes/*kodal*	Cultivators, rotary tillers, cultivators with rigid and spring-loaded tynes
Weeding	Japanese rice weeder/dryland	—
	Weeder, cono weeder/ hoes/ *kodal*/ sprayer, long handled weeders, hoe-cum-rake, wheel hoes	
Plant protection	Manual sprayers, handheld sprayers, backpack and knapsack sprayers, knapsack power sprayer-cum-duster, dusters	—
Foliar sprays	Manual sprayers, manual sprayers, handheld sprayers, backpack and knapsack sprayers, knapsack power sprayer-cum-duster, dusters	—
Harvesting	Sickle/desi plow, paddy reaper	Digger, mowers, self-propelled paddy harvester

Table 6.
Farm types of machinery and implements mostly used by the farmers of the region [5].

- To establish a crop with the optimum plant population during the delayed onset of monsoon.
- To ensure better performance of crops during seasonal drought (early/mid and terminal draught) and extreme events.
- To enhance performance, improve productivity, and income.
- To enhance the adaptive capacity and livelihood of the farmer.

1.12 Crop and cropping system to cope with weather aberration

The whole concept of farming revolves around the seed. Identification of crops and varieties that fit well into changed climatic conditions is a common denominator for sustainable crop production in all land use conditions. An ideal variety should be high yielding, plastic enough to withstand weather aberration, tolerant to multiple abiotic and biotic stresses, responsive to augmented carbon dioxide levels and fit well to farming situations. Sowing the right varieties of the right crops at the right time under the right land use conditions makes a significant difference in attaining higher yields. Efforts have been made by ICAR and State Agricultural Universities to develop high yielding varieties (HYVs) suitable for biotic and abiotic stresses. The short-duration varieties and several climate-ready varieties have already been released to cope with the aberrant climatic conditions.

1.13 Weather aberration during the last decade

Due to aberrant weather conditions, Indian agriculture, in general, is experiencing the delayed onset of monsoon, deviation in rainfall, early season drought, mid-season drought, terminal drought, and flash floods. Most of the studies during the last decades in India have pointed out that India's annual rainfall, together with monsoon rainfall, is trendless and is mainly random in nature over a long period, particularly on an Indian scale. However, large inter-annual and decadal variations have been observed. The summer monsoon rainfall has sown significant decreasing trends in the subdivision of south Assam (−12.5 mm) per decade. The declining trend of annual rainfall at a rate of 7.9 mm/100 years is also reported [10]. The summer monsoon rainfall in the region is decreasing (−48.0 mm/100 years), while rainfall during the post-monsoon season showed an increasing tendency (36.3 mm/1000 years).

1.14 Crop production architecture

Packages for climate neutral variety, flood stress resistant variety, and insect pest resistant/tolerant varieties are to be developed over the years. Science of robotic application in production farm for weed management, etc., and application of GPS in farm planning and soil fertility management will have to be perfected.

Opinions for and against the SRI method of rice cultivation will require to be scientifically assessed to bring in needed refinement for water saving method of rice production. Designing and prototyping suitable farm types of machinery to suit changed crop production planning will need focused attention. A multidisciplinary team involving an agronomist, plant breeder, biotechnologist, meteorologist, entomologist and pathologist, and engineers will have to come together to remodel the crop production module. A team consisting of an expert from different branches of horticulture will have to initiate research to develop smart farming for small farmers, a model of hi-tech/climate control horticulture spreading the dimension to the hydroponic and aeroponic horticulture, including vertical space utilization in the greenhouse and soil-less farming, rooftop farming, *etc*.

The way the rainfall pattern is changing–sometimes early, sometimes late, sometimes scanty, sometimes in excess, sometimes erratic. Therefore, our established method of crop production will demand a change, necessitating advancing sowing time in some cases and delaying the time in some others.

Since 2011, under National Innovations in Climate Resilient Agriculture (NICRA), 23 centers of AICRPDA are conducting on-station and on-farm demonstrations/trials under NICRA with a focus on RTCP implementation and preparedness to cope with weather aberrations. Similarly, alternate and resilient crops and cropping systems are demonstrated in AICRPDA-NICRA villages as preparedness to cope with weather aberration.

2. Adoption of climate-resilient technologies to minimize the effect of climate change in AICRPDA-NICRA adopted villages of Lakhimpur district of Assam – a case study

2.1 Methodology

2.1.1 Climate vulnerability

The climate of the village is characterized by hot and humid summer and dry and cool winter. The village is situated in a high rainfall zone. The long-term average annual rainfall of two nearby stations of the village is 1987 mm (Biswanath Chariali) and 2900 mm (north Lakhimpur) with 125 numbers of rainy days. The rainy season in the village starts in March and the quantum of rainfall as well as the number of rainy days increases gradually and reaches the maximum in July/August and then declines to a minimum during November/December. Temperatures of the village generally remain within a comfortable range, however, when there is a dry spell during the summer season, high temperature along with high humidity increases insect pests and diseases of crops. August is the hottest month and February is the coldest month.

In recent years, a substantial reduction in rainfall amount during monsoon season is noticed in this region. The village experienced drought-like situations in recent years *viz.*, 2001, 2005, 2006, 2009, and 2011. There was a substantial yield reduction of *Sali* rice during those years. In the Lakhimpur district of Assam, there is a reduction in rainfall at the rate of 0.52, 1.86, and 0.24 mm per annum during pre-monsoon, monsoon, and post-monsoon seasons, respectively [9].

Chamua and its adjoining villages have different weather-related problems, such as dry spells, during the growing season of *Sali* rice, scanty rainfall during the *rabi* season, and the occurrence of occasional flash floods in a portion of the village. Almost every year the Ganakdalani village is affected by 3–5 numbers of flash floods of 7 to 15 days duration during *the kharif* season. On the other hand, during *rabi* season due to prolonged dry spells, soil moisture deficit is a problem. Due to the presence of only low-lying land situations, there is limited scope for crop diversification. *Sali* rice grown in the village suffers from floods every year.

2.1.2 Soil type

The altitude of the village varies from 83 m to 90 m, which indicates that the village has different land situations varying from upland to lowland which remains flooded continuously for about 6 to 7 months. The soils in this village are mainly Inceptisols. The soils are of sandy loam to silty clay loamy with pH ranges from 4.7 to 6.4. The organic matter content of the soils of the village varies from 0.34 to 3.03%. The status of available nitrogen (275–540 kgha^{-1}) and potassium (138 to 330 Kgha^{-1}) is medium; however available phosphorus (21.4–54.0 kgha^{-1}) content of the soil is low to medium [20]. High soil acidity, high phosphate fixation, micronutrient deficiency, iron toxicity, periodic soil moisture stress during winter seasons, etc., are some of the soil-related problems of this village.

During 2010–2011, participatory rural appraisal (PRA) has been conducted by a team of scientists of AICRP for Dryland Agriculture, Biswanath Chariali Center with extension experts at Chamua and Ganakdalani village. Following problems have been identified during participatory rural appraisal (PRA).

2.2 Constraints

2.2.1 Weather constraints

Following information about the weather/climate variability in the village was extracted during the PRA study in the village.

- Early season drought/normal onset followed by 15–20 days dry spell, mid-season drought, long dry spell, etc. (2001, 2005, 2006, and 2009).
- Erratic and scanty rainfall during *rabi* season.
- Occasional flash flood due to very heavy rainfall in a portion of the village.
- Groundwater is contaminated with arsenic and problems of iron toxicity.
- High rainfall and inundation of fields at the time of harvest of Ahu rice are limiting factors.

2.2.2 Production constraints

- Small and fragmented land holdings (marginal and small farmers).
- No farm mechanization.
- Upland (25 ha) of the village remains fallow.
- *Sali* rice dominated the rainfed monocropping system with indigenous cultivars.

2.2.3 Constraints related to cultivars

- Farmers grow only a few high-yielding rice varieties, such as *Ranjit and Mahsuri,* on a limited scale.
- Farmers of the village do not grow short or medium-duration cultivars that are suitable for an early vacation of the land for subsequent *rabi* crop.
- Farmers of the village use their seeds year after year.

2.2.4 Constraints related to agronomic practices

- Farmers do not follow scientific water management practices in paddy fields.
- Farmers generally go for haphazard planting or broadcasting instead of planting in rows or line sowing. It has been established that there is a yield reduction of 1.5 to 1.8 tha^{-1} due to haphazard planting.
- No soil test-based nutrients management/INM.

2.2.5 Constraints related to insect-pests and diseases

- Major insect pests of rice in NICRA village are stem borer, gall midge, caseworm, hispa, etc. The major diseases are blast, blight, brown spot, false smut, etc. Farmers of the village do not use any control measures to protect the crop from the insect-pests and diseases. Weeds infestation is a major problem in *Ahu* rice.

2.3 Objectives

To overcome the climatic constraints interventions under National Innovations on Climate Resilient Agriculture (NICRA) under All India Coordinated Research Project for Dryland Agriculture are being implemented under adaptive strategies for adaptation to the climate change effect from 2011 to 2012 to evaluate the technology options for adaptation to the climatic vulnerability through on-farm interventions. The interventions are categorized under two heads viz., real-time contingency planning (RTCP) and preparedness. Under RTCP, interventions are implemented to cope with delayed onset of monsoon, early season/mid-season/terminal drought through the demonstration of proven technologies. As preparedness, interventions are being demonstrated under different themes, such as rainwater management, crops and cropping system, alternate land use, and energy management. With more than 150 participants farmers with the following objectives -.

1. Real-time contingency plan implementation in a participatory mode.

2. Preparedness.

2.3.1 Real-time contingency plan (RTCP) implementation

2.3.1.1 Interventions in case of delayed onset of monsoon

In Assam, the monsoon starts in the first week of June. However, delayed onset of the monsoon was observed in 2011 and 2014 by 15 days and 14 days, respectively, in the village. The villages receive an average rainfall of 400 mm during the pre-monsoon months (March to May). Farmers used to sow the rice seeds in nursery beds depending on the monsoon rainfall. But delay in monsoon affects timely sowing of long-duration rice varieties (more than 150 days) before 15th June. It leads to delay in transplanting and hence affects the production. Farmers were advised to use harvested rainwater in farm ponds during pre-monsoon months to prepare the rice nursery beds in time using water lifting pumps of custom hiring center (CHC). This helped in the timely transplanting of rice seedlings, which facilitated a better establishment, better growth, and better yield of long-duration cultivars as compared to fields of other nearby villages where transplanting was delayed due to delay in sowing. An increase in yield from 12 to 30% has been observed among the rice varieties (*Ranjit, Gitesh, Mahsuri*, etc.) when sown in time (before 15th June) over late sowing conditions (after 20th June).

2.3.1.2 Interventions in case of early season drought

The villages experienced early season droughts of 14 days, 19 days, 11 days, 21 days, and 8 days in 2011, 2014, 2015, 2016, and 2018, respectively. In 2015, in the month of

August, the NICRA village Chamua received 63% less rainfall than a normal one. The rainfall activities in the month were reduced substantially, and the village experienced two dry spells of 6 (10 to 15 August 2015) and 13 (17 to 29 August) days' duration. The effect of dry spell was not prominent in the case of long and medium duration varieties, which were grown in the lowlands, however, short duration cultivars (*Dishang, Luit*) and farmers' varieties, which were cultivated in the uplands of the village and were at early tillering/PI stage affected considerably. Even rice varieties grown in the medium lands (tillering stage) were affected to some extent. The effect of a dry spell on short-duration varieties grown in the medium land situations was comparatively lower than those grown in the upland situation. As the real-time response, farmers of the village were advised to irrigate short-duration cultivars (*Dishang*), which were in tillering or PI stage with supplemental irrigation of 5 cm depth from harvested rainwater. There is an increase in yield of 59% (*Dishang*) over non-irrigated fields.

2.3.1.3 Interventions in case of mid-season drought

The village experienced a mid-season drought of 12 days, 21 days, and 25 days in 2015, 2016, and 2018, respectively. Though the effect of a dry spell on *Sali* paddy was not very much prominent in lowland, early transplanted short-duration varieties grown in the upland situation (active vegetative, early tillering, and PI stage) were affected considerably. During 2016 and 2018, in the medium land situation, long dry spells affected the PI stage of medium duration varieties. In real-time response, farmers were advised to irrigate (2 cm depth) the crop by harvested rainwater in the farm ponds using water lifting pump from the custom hiring center.

2.3.1.4 Interventions in case of terminal drought

The occurrence of terminal drought in long-duration rice varieties in upland and medium land situations has been observed in NICRA adopted villages of the Lakhimpur district of Assam. The terminal drought of a duration of 14, 25, 25, 25, 9, and 25 days during 2011, 2013, 2014, 2015, 2016, and 2017, respectively, have been observed. As a contingency, short and medium-duration rice cultivars were cultivated in upland and mid-land situations, respectively as these varieties can escape mid-season and terminal dry spells. There is an if long duration high yielding and traditional cultivars were cultivated in upland and mid-land situations the crop would have suffered from terminal drought which was experienced by the village. Short-duration cultivars–*Luit, Kolong, Dishang*, and *Lachit*, and medium-duration cultivars–*Shraboni*, *Mohon,* and *Mulagabharu,* were demonstrated as contingency plans for the management of mid-season and terminal droughts. An increase in yield of 26% (*Dishang)*, 17% (*Luit*), 14% (*Shraboni*), and 18% (*Mulagabharu*) has been observed over farmers' practice, *that is*, . cultivation of long-duration cultivars irrespective of the land situation.

2.3.1.5 Flash flood management through varietal manipulation

Unlike NICRA village Chamua, a flash flood is the major weather aberration in the NICRA village Ganakdoloni. Every year the entire village is affected by 3–5 numbers of the flash flood of 7 to 15 days duration, affecting the *Sali* rice grown in that village. In 2012, *Sali* rice grown in the village was under intermittent submergence for a total of 32 days and there was a total crop failure. Different *Sali* rice varieties, including submergence tolerance variety–*Jalkunwari,* grown in the field were damaged. However, one variety–*Jalashree,* withstand the intermittent submergence and was able to give some yield (9.0 qt/ha). In 2013, submergence tolerance (*Jalashree* and *Swana sub-1)*, staggered planting (*Gitesh* and *Prafulla*), and *Bao* rice varieties (*Kekowa, Tulshi, Dhushuri, Bahadur, Maguri*, and *Rangabao*) were in the village. In 2014, locally available six varieties of deepwater rice (*Bao rice*) *viz., Kekowa* (23 qha^{-1}), *Rangabao* (26 qha^{-1}), *Dhushuri* (24 qha^{-1}), *Maguri* (23 qha^{-1}), *Bahadur* (19 qha^{-1}), was demonstrated in farmers' field of Ganakdoloni. Four improved *Bao* cultivars, namely, *Panchanan, Panindra, Basudev,* and *Padmapani* were collected from RARS, Lakhimpur, and cultivated in the village. Despite a very long submergence period (more than 40 days) local *Bao* varieties performed well, however, improved *Bao* varieties were damaged completely by flash floods.

From the experiences gained from the on-farm trials conducted at farmers' fields of the village during 2012–2013, 2013–2014, and 2014–2015, it was found that the normal farmers' varieties, including *Ranjit, Mahsuri, Punjublahi,* etc., could not withstand and were completely damaged. Though submergence tolerant rice varieties can withstand submergence up to 15 days during the seedling and tillering stages, the same varieties fail to survive if exposed to submergence for a few days during or after the panicle initiation stage. Submergence tolerant rice varieties (such as *Jalkunwari and Jalashree)* evaluated in the village did not perform well or exhibited total crop failure in cases when the plants at the panicle initiation (PI) or grain filling stage were exposed to submergence during the latter part of September or in October (as in case of 2012) the flash floods. Some of the improved *Bao rice* varieties were evaluated, which failed to survive during 2013–2014. Against the failure of normal, submergence tolerant improved *Bao rice* varieties evaluated in different crop seasons, six traditional *bao* varieties evaluated in the study area could survive intermittent submergences in both the crop seasons (2013 & 2014). The traditional *bao* rice varieties are having some special characteristics, such as tall stature, elongation ability, and kneeing ability, which are suitable for withstanding short or long-duration submergence.

2.3.1.6 Drought management through nutrient management in Sali paddy

Before the implementation of the NICRA project, farmers of the NICRA village Chamua did not apply any chemical fertilizers in the *Sali* rice growing field. Nutrient management through the balanced application of fertilizers (N, P, and K), and proper method of application proved to be beneficial even in situations like mid-season and terminal drought. Midterm corrections in the case of mid-season and terminal drought were suggested in terms of the application of an additional dose of 22 kg ha^{-1} MOP in maximum tillering to grain growth period to the farmers of NICRA village. An increase in yield of 33% (*Ranjit*), 32% (*Gitesh*), 64% (*Shraboni*), and 57.5% (*Mulagabharu*) has been observed over farmers' practice.

2.3.1.7 Management of drought with alternate crop and crop diversification

Before the implementation of the NICRA project, rice crop was grown on all types of the land situation available (up, medium, and lowland) in the NICRA village Chamua. As rainfed rice requires a higher quantity of water during crop growing season, rice crops are grown in the upland as well as well-drained medium land situations in the village often suffer from a mid-season as well as terminal drought. In a worse situation (experienced in 2006 and 2011), the yield of rice grown in the upland areas of the village even goes below 9 qha^{-1}. Therefore, an alternate strategy of growing high-value crops, such as ginger, turmeric, sesame, black gram, green gram, summer vegetables, and winter vegetables, was taken up as a contingency measure for the management of drought (delayed onset of monsoon, mid-season, and terminal drought). Crop diversification with high-value crops along with organic mulch-cum-manuring proved to be more resilient to stressful situations arising during mid-season and terminal dry spells during *kharif* as well in *rabi* season. Therefore, farmers of NICRA village Chamua were encouraged to take up crop diversification to cope with rainfall variability instead of growing *Sali* rice, especially in upland situations.

2.3.2 Preparedness

2.3.2.1 Rainwater management

Thirteen farm ponds and a canal of 0.5 km length were renovated at *Chamua* village for rainwater harvesting during pre-monsoon and monsoon seasons after the implementation of NICRA. Rainwater harvesting in the renovated farm ponds during pre-monsoon months is used for sowing *Sali* rice in a nursery bed in case of delayed onset of monsoon. Rainwater harvesting in the canal and farm ponds during the monsoon months is efficiently utilized for providing 1–2 supplemental irrigation in *rabi* crops, such as potato and rapeseed. It has also acted as a source of drinking water for the grazing animals during the dry period of the year. Mulching with locally available organic mulch material in ginger, turmeric, potato, tomato, etc. was found more productive than the crops grown without mulching.

2.3.2.2 Crops and cropping system

In upland well-drained loamy soils, short-duration varieties (*Dishang, Luit*) performed consistently better as compared to the farmers' cultivar or long-duration cultivars. It also facilitates the early vacation of the field, thereby helping the farmers to grow *rabi* crops on the same piece of land. In the case of medium land and moderately well-drained soil, medium duration varieties performed better despite the occurrence of terminal dry spells. In lowland situations and soils with poor drainage, the effect of a terminal dry spell is the minimum. Therefore, long duration varieties are grown instead of farmers' variety or medium duration varieties. Efforts are being made to introduce maize in the adopted village. It was observed that this crop can be sown in all types of the land situation available in the village in the driest period (in terms of rainfall received) of the year (December to February) after the harvest of *Sali* rice as well as other *rabi* crops, such as potato and rapeseed. Under NICRA, land situation-specific, profitable, and climate-resilient double cropping systems were identified and implemented in the adopted villages and it was observed that rice equivalent yield, as well as net income, increased significantly in all the identified double cropping systems as compared to the existing monocropping of rice. Growing alternate crops and crop diversification not only helped the farmers to minimize the risk due to extreme weather events but also in better income generation and nutritional security. In flash flood-prone areas, traditional *Bao* rice varieties (*Kekowa, Rangabao, Maguri, etc.*) performed better than the improved submergence tolerant varieties (*Jalashree, Jalkuwari, Padumoni, Panindra*, etc.). It was observed that submergence tolerant varieties could not withstand an intermittent flash flood. Presently, the farmers of *Ganakdalani* village have adopted traditional *Bao* rice varieties and almost 80% of the rice cultivated area of the village is under traditional *Bao* rice.

2.3.2.3 Alternate land use

During 2014–2015, seven low-cost polyhouses were erected in Chamua village under NICRA for demonstration of the cultivation of high-value off-season vegetables and raising vegetable seedlings in advance of *rabi* season. Protected cultivation under low-cost poly houses facilitated the cultivation of high-value crops, such as off-season leafy vegetables, *ghost chilli*, tomato, and cucumber, for better profitability. It was observed that farmers earned on an average Rs. 40,000.00 per year from a low-cost polyhouse of 100 square meters size.

Vermicompost production helped the farmers to meet their organic manure requirement as well as profit from selling the same to other farmers. Average vermicompost production was 1000 kg per tank per annum. From 2018 to 2019, mushroom cultivation was also demonstrated for better nutritional security of the farmers. An average income of Rs.7280.00 has been realized by the farmers as an alternate source of income.

2.3.2.4 Energy management

Since resource-poor farmers cannot afford to purchase the costly farm implements/machines of their own, therefore, custom hiring centers with need-based farm implements/machines (rotavator, cultivator, thresher, reaper, transplanter, water lifting pumps, duster, sprayers, *etc.*) was established under NICRA project for timely completion of farm operations. These implements were made available for hire by the needy at cheaper rates fixed by the NICRA village management committee.

Timely farm operations carried out with the help of the custom hiring center facilitated the farmers to complete sowing or transplanting in time. A study conducted on mechanization revealed that with an increase in the level of mechanization, the human and animal hour requirement for paddy cultivation was reduced from 795 to 350 and 353 to 23 $hrha^{-1}$, respectively. Thus, mechanization helped in a substantial reduction of the drudgery of humans and animals.

2.4 Resource characterization

The total land area under the NICRA village was categorized into upland, medium land, and lowland based on the soil survey report. Soil health cards were also prepared. The presence of a high level of arsenic in groundwater was detected. To minimize the effect of arsenic contamination, a low-cost filter technology (*Arsiron Nilogon*) was demonstrated at the village in collaboration with Central University, Tezpur, Assam.

A mini agromet observatory has been established for the collection of rainfall and temperature data. Dissemination of agromet advisory service to the farmers helped in the decision-making process in the preparedness stage of real-time contingency planning, such as land-related (e.g., land situation wise decision making), rainwater harvesting (mulching, farm pond, micro irrigation system, etc.), crop-related (selection of suitable crop/varieties), and management related (management of insect-pest, diseases, nutrient, weed, etc.).

2.5 Integrated farming system (IFS) is one of the most potent nature-based adaptation strategies

A "whole-farm" approach or integrated farming systems that supplement traditional crops with farming vegetables, fruits, poultry, or fish is re-emerging as a nature-based solution to boosting productivity in climate-stressed regions. A whole-farm approach or pond-based IFS, however, helped increase production, employment, and income by generating a mix of enterprises based on rice, vegetables, fruits, fish, pigs, and poultry. The integrated farming system adopted, tapped into on-farm resources and reduced dependence on external inputs, such as fertilizers and pesticides. The study found that the net returns under an integrated system that included ponds, rice, vegetables, and pigs were eight times higher than the normal farmer practice of growing a single crop. Scientific fish farming and multiple uses of water for growing vegetables, livestock, and irrigation can enhance productivity, income, and employment, as well as resilience against climate change. The integrated farming approach provides year-round production, employment, and income, reducing farming risks during climate uncertainty (**Table** 7) [19].

Adaptation options	Farmers adapted measures
Crop variety change	HYV as well as a hybrid variety of rice, vegetables, jute, maize, wheat, rapeseeds, and mustards
Crop switching/ Mix cropping	1. Mix cropping among winter rice, summer rice, *rabi* vegetables, rape and mustard, jute, wheat, *kharif* vegetables. 2. Crop switching from sugarcane, jute, buckwheat, banana, and lemon to mainly horticultural crops (high-value crops), food crops to non-food crops.
Adjusting planting dates	1. 10–15 days delay in sowing of winter (*Sali*) rice to get the benefits of early monsoon. 2. 15–20 days earlier plantation of summer (*Boro*) rice variety to avoid crop loss due to rain during harvesting time. 3. Delayed planting of *Sali* varieties, such as *Hatisali and Boradhan,* during heavy rainfall. 4. Late plantation of rice, such as *Hira* (*Sali*), in September to avoid heavy monsoon rainfall. 5. Adjusting planting dates of hybrid vegetables, such as turnip, by late planting due to heavy rainfall.
Increase the use of fertilizers	**Rice** 1. N-urea has increased to 20 kgbigha^{-1} from 10 to 15 kgbigha^{-1}earlier. 2. P_2O_5-phosphorous has increased to 25 kgbigha^{-1}from earlier 15–18 kgbigha^{-1}. 3. K_2O-potash increased to 10–12 kgbigha^{-1}from earlier 4–5 kg per hectare. **Vegetables** 1. For *kharif* vegetables, such as tomato, brinjal, and chili urea, is used is 20–22 kgbigha^{-1}, for capsicum, cauliflower cabbage, turnip, carrot it is 25–30 kgbigha^{-1}, broccoli requires up to urea 50 kgbigha^{-1}. 2. P_2O_5-phosphorous used for cabbage, cauliflower, turnip, brinjal, tomato, capsicum, and broccoli are 50–70 kgbigha^{-1}. 3. K_2O-potash requirement is 15–20 kg for cabbage, cauliflower, tomato, turnip, and capsicum. 4. In addition to this well rotten FYM or compost application in nursery beds is used to improve the soil's physical condition. 5. Borax up to 25 kg is used for some vegetables like cauliflower. 6. Organic manures are used up to 2 quintals per bigha by some farmers.

Adaptation options	Farmers adapted measures
Pest and Disease management	1. Farmers use various chemicals, such as *boric acid powder, Bordeaux, ustad, profax, DAP, captan, mancozeb, dithane Z-78, Karathane, bavistin, calixin, and bentate,* to control pest and diseases. 2. After sowing, some local practices, such as covering the seeds with a thin layer of sand mixed with well-dried cow dung, wood fine ash, and dried tree leaves, are mashed and spread in to protect from insects like *thrips*. 3. Dried grass and a banana leaf or thin layers of straw are used in nursery beds of vegetables to prevent displacement of seeds as well as protect from water-borne disease. 4. Burning rice strips and rice plant roots after harvesting to control insects. 5. Seedling tips of rice are trimmed before sowing. 6. Vegetables, such as French beans seeds, are protected by applying black pepper powder to the seeds which prevent storage pests. 7. Applying lime before planting vegetables, such as cauliflower, pea, and carrot.
Using shades or shelter	1. Shading is done by the banana stem to protect the crops from harsh sun rays. 2. Use local plants, such as *Khoria* and *Gancha,* that not only shelters the plants but also help to increase soil fertility.

Table 7.
Documented the following adaptation strategies at the farm level from the field study [5].

The successful interventions under NICRA have also been upscaled across three districts of Assam in collaboration with Krishi Vigyan Kendras and convergence with other state departments, input agencies as well as different financial institutes helped the farmers in terms of technical guidance, agricultural inputs, and financial assistance.

3. Conclusion

Climate change is a complex phenomenon and has many manifestations. It is very difficult to generalize possible remedial measures. An increase in temperate will adversely affect crops by accelerating crop growth rate, reducing crop duration, and crop yield, increasing the rate of evaporation/evapotranspiration, and decreasing fertilizer use efficiency and negative impact on soil health. However, location-specific adaptation and mitigation measures are essential to face the impact of extremes. The social and physical impacts of climate change are heterogeneous as the magnitude and direction of climate change across the globe vary and even within the same regions experiencing climate change are likely to vary [8]. Experiences gained from NICRA implementation showed a changing trend of agricultural practices to cope with changing weather aberrations. The cost-effective and easily adaptable technologies are the backbones for enhancing farmers' income in such situations. Change in variety plays an important role in withstanding crop loss. Likewise, alternate land use systems are very much beneficial for the small and marginal farmers to increase income as well as nutritional security. There is a need for low-cost technologies in the form of variety, small machinery, and other technologies so that farmers could get benefit from the limited resources available to them.

Acknowledgements

The authors express profound gratitude to the Hon'ble Vice Chancellor, AAU, Jorhat, Director of Research (Agri), AAU, Jorhat for their valuable guidance, and professional and administrative support during the implementation of the technical programs. The team is equally grateful to the Director, ICAR-Central Research Institute for Dryland Agriculture, and Project Coordinator (Dryland Research) for financial support and their valuable guidance during the implementation of the project.

Author details

Pallab Kumar Sarma[1*], Nikhilesh Baruah[1], Rupam Borah[1], Rupshree Borah[1], Arunjyoti Sonowal[1], Rekhashree Kalita[1], Prasanta Neog[2], Prabal Saikia[3] and Nipen Gogoi[3]

1 All India Co-ordinated Research Project for Dryland Agriculture, BNCA, AAU, Biswanath Chariali, India

2 Department of Agrometeorology, Biswanath College of Agriculture, AAU, Biswanath Chariali, India

3 Regional Agricultural Research Station, North Lakhimpur, AAU, Lakhimpur, India

*Address all correspondence to: pallab.k.sarmah@aau.ac.in

References

[1] Deka RL, Mahanta C, Pathak H, Nath KK, Das S. Trends and fluctuations of rainfall regime in the Brahmaputra and Barak basins of Assam, India. Theoretical and Applied Climatology. 2013;**114**:61-71

[2] Neog P, Dihingia PC, Sarma PK, Maruthi Sankar GR, Sarmah D, Sarmah MK, et al. Different levels of energy use and corresponding output energy in Paddy cultivation in north Bank plain zone of Assam, India. Indian Journal of Dryland Agricultural Research and Development. 2015;**30**(2):84-92

[3] Bhattacharyya HC, Borkakati K, Ray BK. Agriculture Profile of Assam. In: A Book of Compilation of History, Status, and Strategies of Important Aspects of Agriculture in Assam. Jorhat-Assam: Assam Agricultural University; 2015

[4] Rashtriyariya Krishi Vikash Yojana (RKVY). 2022 http://www.rkvyassam.in/rain

[5] Baruah N, Sarma PK, Borkotoki B, Sonowal A, Kalita R, Neog P, et al. A Book on Improved Agronomic Practices for Rainfed Crops in India. Hyderabad, India: Central Research Institute for Dryland Agriculture; 2022. pp. 46-64

[6] Status Report of National Agricultural Research Project. Assam, North Bank Plains Zone, Status Report. North Lakhimpur: Assam Agricultural University; 1989

[7] Arias PA, Bellouin N, Coppola E, Jones RG, Krinner G, Marotzke J, et al. Technical summary. In: Masson-Delmotte V, Zhai P, Pirani A, Connors SL, Péan C, Berger S, Caud N, Chen Y, Goldfarb L, Gomis MI, Huang M, Leitzell K, Lonnoy E, Matthews JBR, Maycock TK, Waterfield T, Yelekçi O, Yu R, Zhou B, editors. Climate Change 2021: The Physical Science Basis. Contribution of Working Group I to the Sixth Assessment Report of the Intergovernmental Panel on Climate Change. Cambridge University Press, Cambridge; 2021. pp. 33-144

[8] O'Brien KL, Leichenko RM. Double exposure: Assessing the impacts of climate change within the context of economic globalization. Global Environmental Change. 2000;**10**:221-232

[9] Deka RL, Nath KK, Hussain R. Variability in rainfall and temperature regimes of the Brahmaputra Valley of Assam. Journal of Agrometeorology. 2008;**2**:305-308

[10] Deka RL, Mahanta C. Trends in the Rainfall Pattern over North East India. Tezpur, Assam: Tezpur University; 2010

[11] Deka RL, Mahanta C, Pathak H, Nath KK, Das S. Annual and seasonal rainfall trends in the Brahmaputra and Barak basins of Assam, India. Theoretical and Applied Climatology. 2012

[12] Pant GB, Rupa KK. Climate of South Asia. Chichester, UK: John Wiley Sons; 1997

[13] Chaturvedi RK, Joshi J, Jayaraman M, Bala G, Ravindranath NH. Multi-model climate change projections for India under representative concentration pathways. Current Science. 2012;**103**: 91-802

[14] Department of Environment and Forest. Assam State Action Plan on Climate Change. India: Government of Assam; 2020. pp. 1-127

[15] Chary GR, Gopinath KA, Bhaskar S, Prabhakar M, Chaudhari SK, and Narsimlu B. Resilient Crops and Cropping Systems to Cope with Weather Aberrations in Rainfed Agriculture. 2020

[16] Gulati S, Singh M. Energy requirement and management in a potato production system. Potato Journal. 2011;**38**(1):61-66

[17] Srinivasarao C, Dixit S, Srinivas I, Sanjeeva Reddy B, Adake RV, Shailesh B. Operationalization of Custom Hiring Centres on Farm Implements in Hundred Villages in India. Hyderabad, India: Central Research Institute for Dryland Agriculture; 2013. pp. 3-4

[18] Mandal S, Kumar A, Singh RK, Ngachan SV. Road map for farm mechanization in Assam state. Indian Journal of Hill Farming. 2014;**27**(1):51-61

[19] Sikka AK, Kandpal BK, Islam A, Dhyani SK. Climate change and wheat. Geography and You. 2016;**15**(94):53-58

[20] Dutta S, Sarma PK, Neog P, Karmakar RM, Baruah S, Borah P. Soils of NICRA Villages-Survey and Land use planning. AAU/DR/15(BL)/08/2015-16. 2016

Chapter 11

Vulnerability of Climate Change and Potential of Domestic Adaptation and Mitigation Pathways

Abdul Rehman Aslam and Fatima Farooq

Abstract

Pakistan is in the eye of global climate change and among the top five countries which are severely affected by it. Residents facing unprecedented heat waves temperature hikes and flash floods. The vulnerability of climate change varies from sector to sector relied on the adaptive capability and sensitivity of different regions. According to the researcher's knowledge, it is a pioneer study which will broadly disclose the vulnerabilities of climate change in different economic sectors as well as how to cope with it through domestic adaptation and mitigation strategies. Cereals consider a staple food which is most vulnerable to climate change, even in the presence of effective capital and advancement of mechanisation. The core objective is to capture the vulnerability of climate change on cereal production via the Cobb-Douglus Production Function on annual time series data over 1977-2016 with the help of the Liner ARDL-Bound Test Approach. This study encompasses temperature, precipitation and inputs indices as well as greenhouse gas emanations which make it prominent from prior studies. Temperature index and greenhouse gas emanations have been destructive despite the fact precipitation and input indices, cultivated area and rural population have constructive liaison with cereal production over a long period.

Keywords: vulnerability, climate change, adaptation, mitigation, Pakistan

1. Introduction

Since the 20th century, global warming has hastened ominously, due to a large number of greenhouse gas emanations, a fast increasing population and fossil fuel combustion. At a worldwide scale, the occurrence of climate-induced natural disasters increases the frequency and intensity of extreme weather events, floods, droughts, unpredicted hikes in temperature and unseasonal heavy rainfalls. Global climate changes hastily influence all sectors of the economy specifically the agriculture sector through variations in planting patterns and productivity. The frequent appearance of extreme events, incessant increase in greenhouse gas emanations, hike in temperature patterns and sporadic precipitation will pose a great threat to economies. Climate change has a prominent influence on developing countries, especially in South Asia and Africa. Furthermore, arid and semi-arid areas of South Asia are under the significant threat of climate change. Ullah & Takaaki [1] mention that' *Pakistan is listed among hazard-prone countries which are hardly hit by climate change. For the reason that it does not have adequate resources to tackle and recover from several large-scale disasters because it is quite complex and resource-demanding'.* Pakistan is included in the list of those countries that are vulnerable to climate change, and vulnerability is basically because of its geographical, demographical and adverse climate circumstances [2]. Unexpected variations in temperature and precipitation trends as well as frequent floods and droughts increase the vulnerability of residents of the country. *Climate change has increased global mean temperature* [3, 4] *and melting of glaciers and sea* [5], *changes in precipitation patterns, and the rise in sea level* [6]. Pakistan is included in those countries which will bear huge losses by climate change in the future [7].

Pörtner et al. [8] disclosed vulnerability, adaptation and mitigation through words:

> *"The propensity or predisposition to be adversely affected by climate change and human as well as ecosystems vulnerability are interdependent. Furthermore, it encompasses a variety of concepts and elements, including sensitivity or susceptibility to harm and lack of capacity to cope and adapt".*

"Adaptation in human systems, the process of adjustment to actual or expected climate and its effects to moderate harm or take advantage of beneficial opportunities. while in natural systems, the process of adjustment to actual climate and its effects; human intervention may facilitate this".

"Mitigation is achieved by limiting or preventing greenhouse gas emissions and by enhancing activities that remove these gases from the atmosphere".

1.1 Greenhouse gas emission

Greenhouse gases are a combination of CH_4, N_2O and F-gases (HFCs, PFCs and SF_6). Pakistan contributed 0.8% to global greenhouse gas emissions while standing on the 135th rank in per capita greenhouse gas emissions, such as 1.9 tonnes per capita, 1/3 of the world average. It is projected to reduce more than 20% of emissions by 2030. The agriculture sector had got the 2nd number of those contributing 24% in greenhouse gas emissions globally. In the future, GHG emissions levels are expected to increase numerous times in different sectors. It relies on the assumption that the anticipated whole GHG emissions are coherent with the economic growth approach which will be more than triple by 2020 and increase about 23 times by 2050 (compared to 1994 emissions). **Figure 1** illustrates the sector-wise GHG emissions projection during 1994, 2008, 2012, 2015 and 2020, whilst **Figure 2** displays future emission scenarios in 2030 and 2050 in which GHG = greenhouse gas and $MtCO_2$ = million tons of carbon dioxide equivalent.

1.2 Frequency and intensity of temperature, precipitation, humidity, clouds, sunshine and solar radiations

From 1901 to 2000, the annual mean temperature abruptly increased by 0.75°C in the South Asian region, whilst 0.57°C in Pakistan [9]. From 1961 to 2007, the temperature increased by 0.47°C; the year 2004 was recorded as the warmest year until 2007,

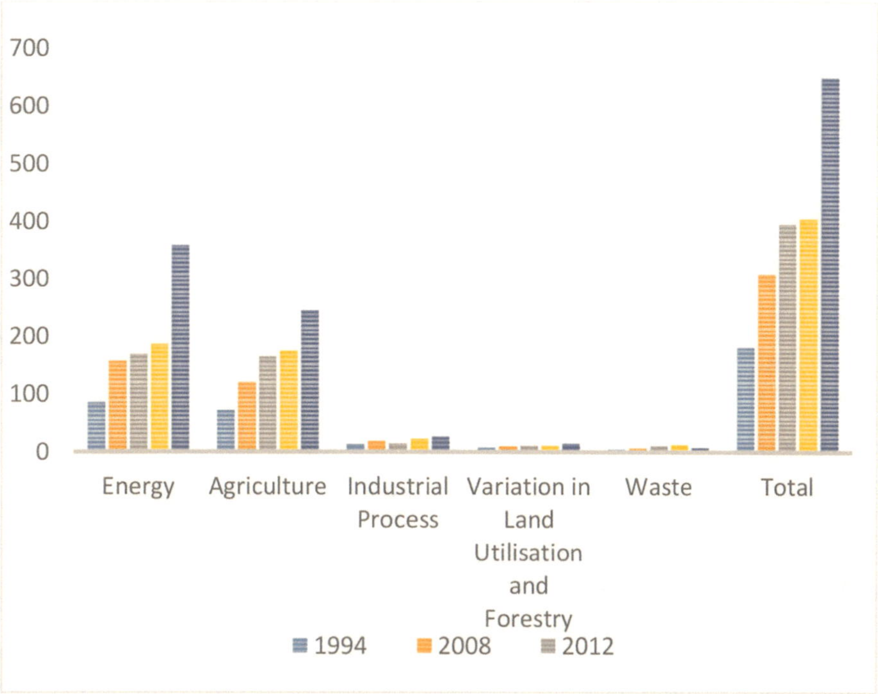

Figure 1.
Sector-wise GHG emissions (Mt CO_2). Source: Author's computations.

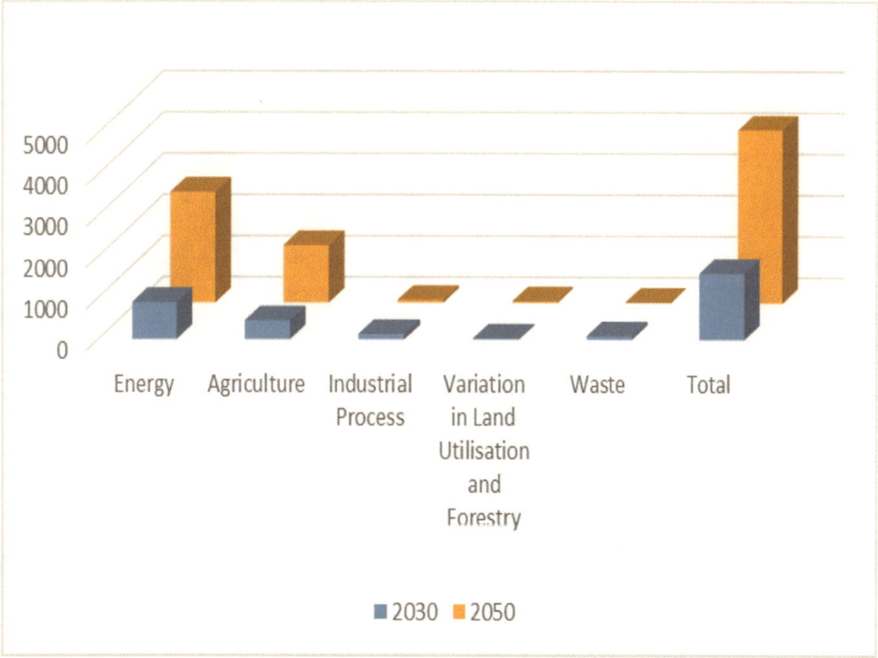

Figure 2.
Future GHG emission scenarios Mt. CO_2. Source: Author's computations.

and the highest temperature rise was recorded during the winter season when it changed from 0.52°C to 1.12°C [10]. During 1951–2000, the mean annual rainfall decreased by 10–15% in arid plains and coastal areas against it increased in the north. During the same time, winter and summer rainfall patterns fell 10–15% in hyper-arid and coastal areas while rose 18–32% in the monsoon zone, particularly in the sub-humid and humid areas. In Baluchistan, relative humidity has recorded a decline of 5%. Furthermore, solar radiation rose by 0.5–0.7% over the southern half, and cloud cover decreased by 3–5% in Pakistan's central region, with an increase in sunshine hours [11].

The various climatic zone-wise annual mean trends of temperature and precipitation are present in **Figure 3**. Zone I (a), Zone III, Zone V (a) and Zone V (b) regions illustrates an increasing trend in the annual mean temperature and precipitation. Zone I (b) and Zone IV regions exposed negative trends in the case of annual mean temperature whilst the adverse trend in the case of annual mean precipitation. Furthermore, only Zone II showed a negative trend in both cases.

1.3 The climate change-induced natural hazards and economic loss

Climate change-induced natural hazards are considered the most devastating reasons for the tremendous loss of human lives, natural resources and infrastructure. According to GCM, climate change-induced hazards (droughts, cyclones, floods, landslides, heat and cold waves) would turn into a greater magnitude, and intensity, more frequently in different regions of Pakistan. The last decade of the 20th century in the Bay of Bengal and the Arabian Sea recorded cyclones, storms and depressions

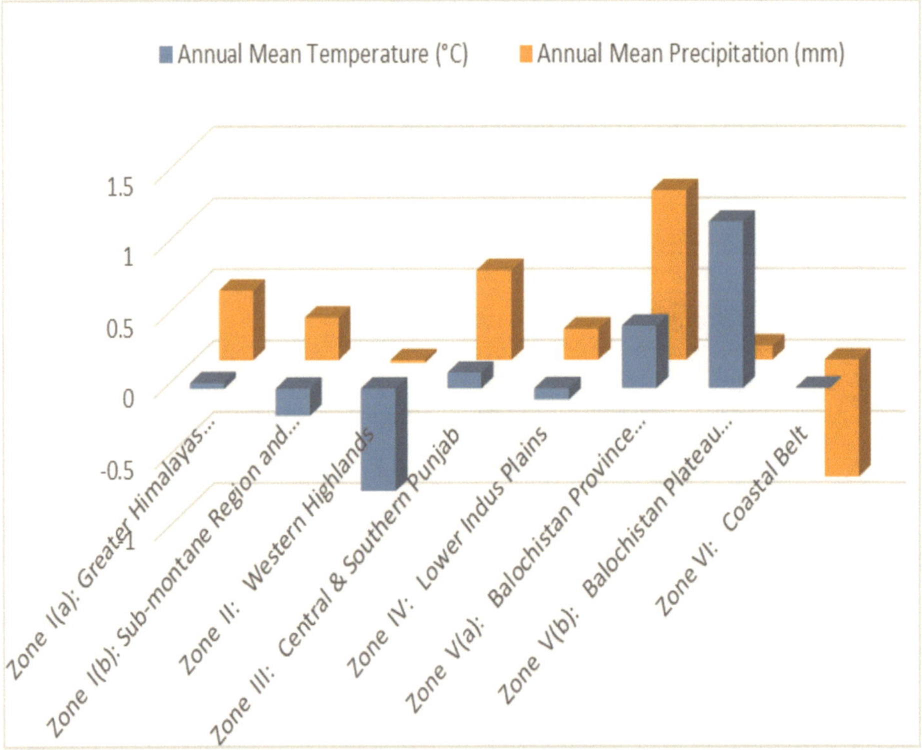

Figure 3.
Zone-wise temperature and precipitation trends (1951–2000). Source: Author's computations.

with high frequency that hit Pakistan and other linked countries of it. Furthermore, this century has also faced seven strong El Niño events. Pakistan's coastal communities of Baluchistan and Sindh provinces are vastly vulnerable to cyclones: Gwadar, Awaran, Lasbella and Ketch and Badin, Karachi, Hyderabad and Thatta, respectively. (GoP[a]) mention that from 1971 through 2001, 14 cyclones were noted. During 1994–2013, extreme climate events did an average annual economic loss of US$ 4 billion. The Yemyin cyclone (2007) hit coastal communities, and about 1000 people lost their lives; above 2 million people were affected by power outages and water shortages, and the cyclone killed more than 2 million livestock. The predicted losses of property from the storm were around 24 billion rupees.

In Sindh and Baluchistan, regions frequently faced droughts because of the decrease in rainfall patterns and other socio-economic situations. The worst drought of history (1998–2001) in South Asia special in Pakistan allied with the La Nina phenomena [12]. Larsen et al. [13] point out that droughts in the period of 2000 and 2002 disturbed more than 3 million people and loss millions of livestock nearly US$ 2.5 billion. According to Government of Pakistan [14] in 2015, hundreds of people lost their lives due to the drought in Thar.

During 1950, 1973, 1976, 1992, 2010 and 2022, major floods were recorded with massive losses of human beings, and the agriculture sector, infrastructure and economy bore losses of near about US$ 400 billion. The floods of 2010–2014 affected 38.12 million people, damaging 3.45 million houses, destroying nearly about 10.63 million acres of crops destroyed and causing more than US $18 billion in monetary losses [15]. The flash flood of 2022 has had devasting effects on human and livestock casualties in addition to extensive home and infrastructure destruction, such as 1600 people killed and 12,900 injured, 805 thousand houses destroyed and 1.2 million houses damaged, 1.1 million livestock lost as well as 13.1 thousand kilometres of roads damaged [16]. Furthermore, the economic cost of the recent severe flood is US$30 billion [17].

A heatwave is defined as a constant increase in maximum temperature intended for a particular time frame, whilst in the case of a cold wave, it will be the opposite. Zahid & Rasul [18] disclosed that '*during 1961–the 1990s, the heatwave events were not as much of common, but throughout 1990-2018s, a rise in frequency is observed*'. Khan et al. [19] exposed that Pakistan faced intense heatwave events in the southwest and the southeast regions due to the high population rate. The month of March 2022 was recorded as the warmest month in the last 60 years; in addition, March to June experienced brutal heatwaves and the hottest temperatures especially in the Nawab Shah city of Sindh, which recorded 51°C and is situated in the southern part of Pakistan. Across Pakistan especially in Sindh, thousands of people were killed due to unprecedented heatwaves in 1952, 1978, 1984, 1988, 2002, 2006, 2009, 2010, 2012, 2013, 2015, 2016, 2017, 2018 and 2022. North-western areas of Pakistan have experienced cold waves, which rise for 30–60 days; on the other hand, in the case of Punjab and the southern areas of Sindh, their trend declined. The average number of cold wave days did not show any significant trend across the country.

Smog is an amalgamation of smoke and fog, and it is described as a combination of several gases with dust and water vapour. It occurs mostly in dry and cold winter months. Three to 4 years ago, smog hit Pakistan, particularly Punjab; it caused lung diseases and breathing problems. Pakistan's northern areas, particularly Azad Jammu Kashmir (AJK), were affected by frequent landslides. Larsen et al. [13] explored that Azad Kashmir, the northern Areas and Khyber Pakhtunkhwa (KPK) were vulnerable to landslides. Since 1926, 13 landslide events have occurred and 413 lives lost, and avalanches have also occurred in the Gilgit Baltistan region, northern parts of KPK and Azad Kashmir.

1.4 Sea level and coastal areas

Climate change threats to coastal areas including Karachi may increase cyclonic activity and sea-level rise due to a higher sea surface temperature. Chaudhry [10] mentioned that *'from 1856 to 2000 coast sea level rose at 1.1 mm per year (mm/year)'*, whilst Pakistan was the least affected by it. Khan & Rabbani [20] point out that *'Pakistan has 1,046 km-long coastlines with the Arabian Sea in the South falling within Sindh and Balochistan. The coastal zone of Sindh vulnerability is greater than the Balochistan despite its flat tidal topography and higher speed population growth due to coastal areas and industrial activities for instance Karachi A 2-m SLR area of 7500 Km^2 probably goes under the Indus Delta and Pasni, can also be affected by SLR since 1.4 mm mean sea level in the coastal areas. On the other hand, through the Indian Ocean plate, the Balochistan coast is tectonically active in addition to uplift at the rate of 1–2 mm per year'*. The sea level has remained all but level since 1960, which is a 60-year period where nothing has happened to the sea level despite the purported steady increase in global temperature over the same period. In the future, if the sea level increases, it will raise the erosion rate beside the coastal belt, decline the quality of drinking water, degrade mangrove forests and cause a decline in fish and shrimp production. Coastal erosion and monsoon waves are also a threat to agriculture farms, which are the source of food and fuel for the population located along the west Makran coast. Pakistan's mangroves are assessed to be declining by 4900 hectares per year, which is alarming [21]. These are being cut to construct shrimp farms and other coastal developments. Near about 80% of fish caught from the coastal water are used for the food web [22].

1.5 Water resources

Water resources are considered as the utmost sensitive sectors of climate change; it also negatively influences water availability and quality. *'Pakistan's Indus Basin Irrigation System is known as the world's largest contiguous system that is mainly reliant on precipitation, snowmelt glaciers, and groundwater abstraction'* [10]. The temperature rise worst affects glaciers and snowfall, which lead to increased stream flows beside glacial lake outburst floods in the coming decades. Snow cover has decreased by about 10%, whilst per capita water availability is down from 5000 of water to 1000 m^3/capita and is projected to decrease from 1000 m^3 to 800 m^3 by way of 2025, transitioning it from a water-stressed to a water-scarce country. Chaudhry [10] highlighted *'the use of water in different sectors such as agriculture (92%), industries (3%), as well as domestic and infrastructure (5%)'*. In the future, sectoral demand for water will rise because of the high rate of population and socio-economic development.

1.6 Agriculture sector

The agriculture sector is considered the most sensitive to climate change. It can influence food quality, accessibility and availability. Agricultural cultivation through irrigation and spate farming systems is vastly sensitive to temperature and precipitation trend variability, and a decline in water availability may lead to a reduction in productivity. It plays an essential role in the economy as it adds 19.2% to the gross domestic product (GDP) and engages 38.5% of the labour force. Almost 65–70% population are directly and indirectly joined with it [23]. GCISC has highlighted that during Rabi and Kharif seasons, the increase in average maximum temperature is projected such as 1–2.0°C for RCP 4.5 and 5–6°C for RCP 8.5, which have serious implications for the agriculture sector [23]. Dehlavi et al. [24] evaluated that increase in temperature (+0.5–2°C) would lead to a decline in agricultural productivity by 8–10% by 2040. Athar [25] found out that major crops yield to fall, specifically wheat and rice and the growing season's length in four agro-climatic zones.

Chaudhry [10] mentioned that the findings of the World Bank Knowledge Portal in these words all major crops and cereals yield will decline, while the yield of wheat will reduce very sharply at the end of 2080 as well as the yield variations, specifically in case of wheat production is panic intended for Pakistan, furthermore significant adaptation and mitigation strategies requires to tackle it. Chaudhry [10] highlighted that the temperature trend increased at night time greater than in the daytime. It showed negative effects on agricultural productivity in heat stress, increased water requirements, and higher respiration rates. Besides the temperature rise, extreme events contribute to losses of crops and land productivity, pest attack, and soil erosion. The warmer climate may bring minor yield improvements in the north by extending the growing period, while staple crop yields (wheat, rice and maize) mostly grown in the south are predicted to decrease.

1.7 Livestock

Livestock is an imperative part of agriculture and plays a vital character in the rural economy, especially in socio-economic development. It contributed 11.53% of GDP and 60.07% in value-added in agriculture, and 8 million rural families are associated with it [23]. Near about 8 million folks are involved and receive more than 35–40% income from it. Ahmad et al. [26] described that a wide range of pastures and rangelands are livestock sources. It is projected that 60% of the land be utilised as rangeland in arid, semiarid areas of Punjab and Sindh, northern areas and Balochistan. Mansoor et al. [27] mention that just in Balochistan, rangelands provide feed to near about 93 million livestock which is nearly 87% of the whole population originating their means of support from it.

Mir and Ijaz [28] mention that GHG emissions from livestock are considered a large part of the agriculture sector's total emissions. For instance, approximately 40% of total GHG emissions while 90% of the agricultural sector emits from fermentation and manure management. It predicted that the impacts of climate change on livestock would be in the form of heat stress and drought leading to a reduction in animal productivity, degradation of grazing systems, a decline in quality and quantity fodder production in addition to developed stress on rangelands and water resources as well as epidemics of diseases.

1.8 Fisheries

Climate change also hazards marine and aquaculture fisheries. It is the core resource of diet employment and income generation in coastal communities. According to Government of Pakistan, [23, 29] GDP share is 0.4%, while overall marine and inland fish production remained at 690.600 thousand m. tons in addition to the exported value of US$ 303.606 million. The Indus Delta's rise in temperature may lead to a drop in river flows and superior water habitat and species loss. Furthermore, the surface water of mountainous regions and the Arabian Sea felt an increase in temperature which adversely affects the growth, existence, reproduction and migration of cold water species.

1.9 Energy sector

This sector has causes and effects of climate change such as rising population, economic growth, varying consumption patterns and the summer month's rising demand for air conditioning which will likely increase energy demand and consequently increase GHG emissions from the energy sector in Pakistan. During 1990–2016, electricity accessibility rose from 59.6 to 99.15%. Unfortunately, continuing power outages in the summer season because of the demand for cooling on top has crippled the economy with an annual loss of 2.6% of $5.8 billion. A comparison of the fiscal year 2012–2013 and 2022 contribution of energy sources was: thermal energy contribution declined from (64%) to (60.9%), followed by nuclear and renewable energy demonstrating an increasing rate from 5 to 12.4% and 0 to 3%, respectively, while hydel energy showed decreasing rate from 31 to 23.7 energy. During 2021–2022, the electricity consumption pattern indicated that the household sector is the largest consuming sector (47%), followed by industries (28%), the agriculture sector (9%) and commercial consumers (7%) in addition to other sectors (8%). Successive governments have affianced National Electricity Plans which are related to future power generation projects, pricing issues and setting high standards for power consumers.

Climate change showed a solid impact on power generation through asymmetrical river flows, concentrated and numerous floods in addition to droughts. Hydel power contributed 23.7% to the energy supply. However, the rise in temperature and fall in rainfall patterns will decline water supplies, leading to a shortfall in energy production and creating an alarming situation for a future scenario. Hydel power production is determined in the north, where the temperature rise is predicted to be at the highest level. In the future, climate change-persuaded natural threats might negatively influence oil besides gas infrastructure because immense precipitation leads to a flood. Most gas fields are in the neighbourhood of the Indus River and are near about 1/3 of primary commercial energy requirements satisfy through imported oil transported by sea mode. Any impairment in infrastructure might break the supply intended for a long period and create a large burden on the national economy. Higher temperatures will increase the evapotranspiration rate, leading to rising electricity requirements for pumping water used for agricultural irrigation [10].

1.10 Human health

Climate change can influence health and environmental factors such as clean air, safe drinking water, sufficient food and secure shelter. It may be over and done with extreme heat events, variable rainfall patterns and natural disasters. The fluctuations in temperature and rainfall patterns are associated with the spread of different infectious diseases and food security [30]. Furthermore, extreme events are also linked with the inhabitant's mental health, such as depression, aggression and distress. Rise in temperature increases the risk of water-borne and vector-borne diseases. Khalid and Ghaffar [31] disclosed that climate change is projected to increase malaria and dengue by 12–27% and 31–47%.

1.11 Deforestation and biodiversity

Forests play a decisive role in ensuring soil and water preservation and decreasing climate change's destructive effects. One of the major reasons for vulnerability to climate impacts is a very low forest cover area. *"Pakistan is included in the list of low forest cover countries with only 5% of land area under forest and tree cover; however, the international requirement is 25%"* [2]. The average deforestation rate is 27,000 hectares per year of which 84% of cut wood is utilised for domestic purposes [2]. Forests are known as a crucial natural resource for the livelihood of rural areas. It makes available timber, food, fuelwood, habitat designed for wildlife and numerous significant ecosystem facilities, for instance reducing or controlling cyclones and storms in coastal areas and mitigating carbon dioxide. *"It is projected that the impacts of climate change, for instance, variation in temperature and precipitation pattern, increasing intensity and frequency of extreme events will sharply affect the forest in addition to threatening the biodiversity status, and soil quality"* [10]. Trios et al. [32] explored that "At the end of the century climate change is turning into one noteworthy driver of biodiversity loss". Climate change would also worsen the hazard of biodiversity due to variations in land utilisation besides population pressure.

1.12 Population growth and infrastructure

Pakistan consists of 5th most populated country and top 10th largest labour force in the world. Its estimated population is 215.25 million folks, while at the time of independence in 1947, they were only 32.5 million [23]. According to 1951, 1961, 1972, 81 and 1998 censuses, Pakistan had a population of 33.7 million, 42.8 million, 65.3 million, 83.783 million and 132.35 million, while on the 2017 census, the population of Pakistan was 207.77 million [33]. The average annual population growth rate is 1.80% which shows that rapid population increases refer to a rise in demand for water and agricultural products. The population of 10 major cities increased by 74.8%; major reason for urbanisation is the accessibility of better socio-economic facilities for mass. Chaudhry [10] revealed that climate change influences urban agglomeration is well-thought-out due to variations in weather or climate change in either duration or magnitude. From past experiences, it is explored that in Pakistan, most infrastructure is situated in hazardous areas from variations in climate. Generally, the urban infrastructure facilities are mutually dependent; a let-down of any single infrastructure leads to distractions into further associated urban facilities. In the next coming decades, climate variation might escalate the frequency of disturbances.

A review of empirical studies was conducted to access the influence of climate change on the agriculture sector. Verge and De Kimpe [34] studied the liaison between greenhouse gases emission and crop production and pointed out that Asia will be a gigantic food consumer and greenhouse gas producer. Gowdy [35] rendered a new method to analyse climate change and economic development through a pragmatic approach. South Asia released a massive quantity of carbon dioxide. Ahmed and Schmitz [36] investigated the economic impact of climate change on the agriculture sector in Pakistan and accessed that climate change adversely affects it. Siddiqui et al. [37] evaluated that increase in temperature in the long and short run was beneficial and harmful, respectively, for wheat production, while the increase in precipitation was beneficial. Variation in temperature besides precipitation is favourable for rice production but the opposite for cotton crops. Furthermore, the temperature rise is harmful to sugarcane production. Janjua and Samad [38] explored that climate change will not destruct Pakistan's wheat production over in long and short periods. Zhai et al. [39] found out technological progress induced while climate variables had detrimental impact on wheat yield in China. Simionescu et al. [40] explored the greenhouse gases effect in the EU and found out a positive relationship among them. Chandio et al. [41] examined that carbon dioxide has a positive influence in the long and short-run, while temperature and rainfall negatively influence just in long run in the case of China. Ahsan et al. [42] quantified a positive relationship among temperature, rainfall, carbon dioxide and cereal crop production in the case of Pakistan. Chandio et al. [43] investigated that temperature and carbon dioxide have a negative association with cereal yield while contrasting in rainfall.

Warsame et al. [44] revealed that rainfall and temperature have positive and negative while carbon dioxide has no significant effect on crop production. Abbas [45] evaluated the impact of climate change on wheat and maize crop production in Pakistan with the help of a panel-pooled mean group. Temperature revealed significant negative results on crop production during the long run while insignificant in the short run. Arable land and fertilisers gave positive and significant effects in the long and short run while improve quality seeds showed the insignificant result. Baig et al. [46] accessed the asymmetrical effects of climate change on rice production in India. This research finds out that rice production has a negative relationship with mean temperatures in the long run while positive in the short run. Positive shocks in rainfall and CO_2 emission have a significantly negative effect in the long and short run. Furthermore, cultivated area, agriculture credit and fertiliser consumption positively affect rice production. Chandio et al. [47] probed the impacts of climate change on cereal production in Bangladesh. CO_2 and rainfall have a negative and positive impact on cereal production in the long and short run, respectively, while temperature has an adverse impact just in the short run. Furthermore, arable land, energy consumption and financial development have positive impacts on cereal production under both runs.

Furthermore, a considerable number of researchers had provided literature on the vulnerability of climate change and its impacts on the rest of the world at different levels and sectors for instance [48–62]. Specific vulnerability-related literature gave attention to systems, places and activities [63, 64], whereas others focus on ecosystems, livelihoods, individuals and landscapes [65].

Pakistan has high-altitude mountains from north to west, arid deserts in the south, the hot and dry Indus River Valley in the centre to the south and a humid 990 km coastline. In this passing century, climate change has become a massive challenge because of its geographical location and its reliance on natural resources; further pressure on water resources, the agriculture sector, human health, and the energy sector, so it has a long-term catastrophic effect on Pakistan's economy. According to the researcher's knowledge and literature review, no sole prior study employs temperature, precipitation and input indices in addition to greenhouse gas emission against these drawbacks; this research will overcome the gap and give a new direction to researchers, policymakers, and national and international organisations. Moreover, the core objective is to evaluate the vulnerability of climate change on cereal production in Pakistan and based on the outcome provide adaptation and mitigation strategies to tackle climate change. On these accounts, it will increase the scope of the study and the nature of the analysis. The research is planned in the following channel; this first section introduction consists of 12 subsections which comprehensively discuss all related aspects and literature review of prior studies; the second section deals with conceptual framework; the succeeding third and fourth sections deal with data collection, model and methodology as well as analysis result with discussion. The last fifth section contains the conclusion, policy implications and a brief history of legislation and plans in Pakistan.

2. Conceptual framework

The liaison between climate change and cereal production reconnoitre through three approaches: Production Function approach (agronomic models)[1], Ricardian approach (hedonic models)[2]and Simulation models. Firstly, the inkling of climate change and agriculture acquaint [66–68] by exhausting "agronomic models" to evaluate the environmental variable's impact on agriculture crop yield. Charles Cobb and Paul Douglas developed and tested the Cobb–Douglas in contrast to statistical evidence during 1927–1947. Afzal et al. [69] mentioned the pros of Cobb Douglas Production Function; it is more appropriate when observations are not large, and it is easy to estimate and interpret. Production theory assumes that the liaison among multiple outputs and multiple inputs is replicated through the concept of transformation function [70] further assumptions (e.g., free disposal of inputs and exclusion of technical inefficiency) in addition to aggregation of all outputs, the input-output liaison is often reduced to a production function) [71] in which one output depends on multiple inputs: Y = f(x). In the traditional specification of the production function, all inputs are treated symmetrically; that is, they are assumed to contribute to the output in the same way [70].

3. Method (data collection and analysis)

This study encompasses annual time series data of Pakistan over 1977–2016 in details temperature, greenhouse gas emanations, precipitation, cereal production, cultivated area and rural population data acquired from World Development Indicators, while seeds availability, water accessibility, fertilisers, mechanisation and agriculture credit comes from Economic Surveys of Pakistan. Prior studies employed temperature and precipitation data (maximum or minimum or average) which are not favourable for cereal production; these models may have either under- or overestimated; in addition to wind speed, sunshine and humidity also play a core role in cereal production but due to unavailability of data going to skip them. Furthermore, this current research developed temperature and precipitation indices which are the summation of maximum, minimum and average data, while the input index is a combination of water accessibility, seeds availability, fertilisers, agriculture credit and mechanisation; in addition, all indices are constructed by principal component analysis. The rural population engages as a proxy of the rural labour force due to the inability of data. Furthermore, variable codes with descriptions and measuring units are listed in the **Table 1**.

The prior studies related to the impact of climate change on cereal [72] in the case of Burkina Faso, [73, 74] empirical evidence from China also adapted Cobb–Douglas production function for estimation purpose. Now, it is also extended according to the present study in which α_0 is intercept, and $\alpha_1 \ldots \ldots \alpha_5$ are parameters of variables:

$$\mathrm{CP_t} = \alpha_0 + TI_t^{\alpha_1} + PI_t^{\alpha_2} + GG_t^{\alpha_3} + CA_t^{\alpha_4} + II_t^{\alpha_5} + RP_t^{\alpha_6} + \varepsilon_t \quad (1)$$

---

[1] *The production function approach is based on the agronomic models that consist of mostly controlled experimental studies.*

[2] *These models are often based on cross-sectional data and based on the Ricardian approach.*

Variables	Description	Measuring Units
CP (Predictand)	Cereal Crops Production	Kg per Hectare
TI	Temperature Index	Celsius
PI	Precipitation Index	Millimetres
GG	Greenhouse Gases Emanation	kt of CO_2 equivalent
CA	Cultivated Area	Hectares
II	Input Index	
RP	Rural Population	Million
Source: Author's Computations.		

Table 1.
Data description.

This algebraic form of the Cobb–Douglas production function changed into log form:

$$\ln\ \mathrm{CP_t} = \alpha_0 + \alpha_1 \ln TI_t + \alpha_2 \ln PI_t + \alpha_3 \ln GG_t + \alpha_4 \ln CA_t + \alpha_5 \ln II_t + \alpha_6 \ln RP_t + \varepsilon_t \quad (2)$$

This study takes one ARDL approach introduced by Pesaran and Shin [75] and Pesaran et al. [76] to search out the cointegration liaison among variables when their integration order is I(0), I(1) or assortment. It is statistically significant for small data span, restraint to employ in same order integration and provide unbiased outcomes in the amalgamation of I(0) and I(1). There is no necessary and sufficient condition to find out the stationarity order of integration, but to keep a safe path and avoid I(2), search out variables stationarity through ADF[3]test; on the bases of its outcomes, move towards L-ARDL[4] technique with the Bound test. For instance, some previous empirical studies [42, 44, 47, 74, 77] also had employed the same techniques to determine the impact of climate change on cereal production. The following model of ARDL is employed for cointegration testing among predictand and regressors:

$$\begin{aligned}\Delta \ln CP_t = \Omega_0 &+ \sum_{a=1}^{s} \Omega_{1a} \Delta \ln CP_{t-a} + \sum_{a=1}^{r_1} \Omega_{2a} \Delta TI_{t-a} + \sum_{a=1}^{r_2} \Omega_{3a} \Delta PI_{t-a} \\ &+ \sum_{a=1}^{r_3} \Omega_{4a} \Delta GG_{t-a} + \sum_{a=1}^{r_4} \Omega_{5a} \Delta \ln CA_{t-a} + \sum_{a=1}^{r_5} \Omega_{6a} \Delta \ln II_{t-a} \\ &+ \sum_{a=1}^{r_6} \Omega_{7a} \Delta \ln RP_{t-a} + \Psi_{1a} \ln CP_{t-a} + \Psi_{2a} TI_{t-a} + \Psi_{3a} PI_{t-a} \\ &+ \Psi_{4a} GG_{t-a} + \Psi_{5a} \ln CA_{t-a} + \Psi_{6a} \ln II_{t-a} + \Psi_{7a} \ln RP_{t-a} + \phi_t\end{aligned} \quad (3)$$

3 *Augmented Dicky Fuller*

4 *Linear Autoregressive Distributed Lag*

Here, Δ is a sign of the first difference and ϕ error term. The null hypothesis (no cointegration) is articulated as:

$$H_0 : \Psi_1 = \Psi_2 = \Psi_3 = \Psi_4 = \Psi_5 = \Psi_6 = \Psi_7 = 0.$$
$$H_1 : \Psi_1 = \Psi_2 = \Psi_3 = \Psi_4 = \Psi_5 = \Psi_6 = \Psi_7 \neq 0.$$

OLS[5] is the methodology proposed for testing long-run liaison among point-out variables in addition to F-statistics employed for the testing hypothesis that indicates cointegration prevails or not among prediction and regressors. The decision is made on lower and upper bounds values: if the statistical value is higher than the value of the upper bound, then reject H_0 (no-cointegration) otherwise case will be the opposite case if values rely on among both bounds then the result will appear as inclusive. In the case of cointegration, move on to the second step which highlights the long-run relationship among variables. The long interval of time for cereals production model (s, r_i) undertakes the succeeding form:

$$\begin{aligned}\Delta \ln CP_t = \Omega_0 &+ \sum_{a=1}^{r} \Psi_{1a} \ln CP_{t-a} + \sum_{a=1}^{r_1} \Psi_{2a} TI_{t-a} + \sum_{a=1}^{r_2} \Psi_{3a} PI_{t-a} \\ &+ \sum_{a=1}^{r_3} \Psi_{4a} GG_{t-a} + \sum_{a=1}^{r_4} \Psi_{5a} \ln CA_{t-a} + \sum_{a=1}^{r_5} \Psi_{6a} \ln II_{t-a} \\ &+ \sum_{a=1}^{r_6} \Psi_{7a} \ln RP_{t-a} + \phi_t\end{aligned} \quad (4)$$

Here, Ψ and ϕ_t are the long intervals of time parameter and error term, correspondingly. The last step is to insist on an error correction model for evaluating the dynamic in cereals production in a short interval of time specified as:

$$\begin{aligned}\Delta \ln CP_t = \Omega_0 &+ \sum_{a=1}^{r} \Omega_{1a} \Delta \ln CP_{t-a} + \sum_{a=1}^{r_1} \Omega_{2a} \Delta TI_{t-a} + \sum_{a=1}^{r_2} \Omega_{3a} \Delta PI_{t-a} \\ &+ \sum_{a=1}^{r_3} \Omega_{4a} \Delta GG_{t-a} + \sum_{a=1}^{r_4} \Omega_{5a} \Delta \ln CA_{t-a} + \sum_{a=1}^{r_5} \Omega_{6a} \Delta \ln NCI_{t-a} \\ &+ \sum_{a=1}^{r_6} \Omega_{7a} \Delta \ln IPC_{t-a} + \omega ECM_{t-1} + \phi_t\end{aligned} \quad (5)$$

Here, Ω and ϕ_t are the short intervals of the time parameter and error term, respectively.

4. Analysis results

In the case of L-ARDL, there is no necessary and sufficient condition to check the stationarity of the variables although this research did this in the first step to keep the safe side and confirm that no single variable is stationary at 2nd degree. Results in the **Table 2** indicate that all the variables are stationary at the level except cereal

5 *Ordinary Least Squares*

Variables	Intercept	Intercept and Trend	None	Outcomes
CP	8.786	8.678	6.704	I(I)
TI	8.203	8.087	8.282	I(0)
PI	13.361	13.182	13.540	I(0)
GG	4.750	4.670	1.349	I(I)
CA	5.370	5.527	2.866	I(I)
II	4.846	4.840	3.230	I(0)
RP	6.550	6.460	2.082	I(0)

Source: Author's Computations.

Table 2.
Outcome of ADF test on level.

production, greenhouse gases emanation and land under cereal production on the level at 5%, so now, it is confirmed that the L-ARDL is the best option and Akaike information criterion chose for optimum lag order. Bound test outcomes and long- as well as short-run outcomes are mentioned in **Tables 3** and **4**, respectively.

The estimated F $\ln CP_t$ ($\ln CP_t/TI_t$; PI_t; GG_t; $\ln CA_t$; $\ln II_t$; lnRPt) came out 5.840 and superior from upper and lower analytical outcomes of the bound test at 5 and 10% outcomes presented in **Table 2**. Furthermore, Eq. (6) and **Figure 4** disclose the cointegration among predictand and regressors during 1977–2016.

$$\text{Cointeq} = \text{CPG} - \left(-1.000^* \text{TI} + 0.175^* \text{PI} - 0.617^* \text{GG} + 2.384^* \text{CA} + 0.303^* \text{II} + 0.095^* \text{RP} \right) \quad (6)$$

The empirical outcomes of the long and short run are presented in **Table 3** panel A. Temperature index has a coefficient value of −1.000 and a significant probability value of 0.029, indicating that cereal production will drop when the temperature upsurges. The previous studies outcomes are sustained by Pickson et al. [74], Chandio et al. [41], Chandio et al. [43], Warsame et al. [44], Chandio et al. [78] and Baig et al. [46] in case of India by using N-ARDL on data from 1991 to 2018 found that temperature has an adverse influence on rice production. In a short interval of time, the coefficient of temperature index is 1.049 and significant which means that a 1°C increase in temperature will increase the 1.049 tonnes of cereal production which is reinforced by Chandio et al. [79] and Baig et al. [46]. While contradicts [79] attained on the rice crop of Pakistan coving 1968–2014 through ARDL and [47] relied on ARDL achieved in cereal production in Bangladesh during 1988–2014.

The precipitation index has a coefficient value of 0.175 and a significant probability value of 0.022 which illustrates that cereal production will trigger as precipitation increases. This outcome is congruent with previous results [43, 44, 47, 72, 74, 78, 80, 81] achieved on cereal production in Bangladesh. The short-run coefficient of the precipitation index is significant, and 0.092 reveals that a 1 mm upsurge in precipitation will lead to a 0.092 tonnes escalation in cereal production [43, 47, 72, 74, 80, 81] also sustained the same results.

$lnCP_t$ = f ($lnCP_t$/ TI_t;PI_t;GG_t; $lnCA_t$; $lnII_t$; $lnRP_t$)			
F-Statistics 5.840			
Critical Values			
5%		10%	
I(0) Bound	I(1) Bound	I(0) Bound	I(1) Bound
2.04	3.24	1.75	2.87
Source: Author's Computations.			

Table 3.
Outcomes of bound test.

Predictand: ln CP; Selected Model: ARDL (1,1,0,1,1,0,0)				
Predictors	**Coefficient**	**Std.Error**	**TStat.**	**Prob.**
Panel A: long and short-run outcomes				
TI	−1.000	0.437	−2.285	0.029
DTI	1.049	0.420	2.498	0.018
PI	0.175	0.072	2.416	0.022
DPI	0.092	0.043	2.150	0.040
GG	−0.617	0.187	−3.284	0.002
DGG	−0.153	0.220	−0.696	0.492
LnCA	2.384	0.944	2.524	0.017
D(lnCA)	1.365	0.566	2.410	0.022
LnII	0.303	0.118	2.558	0.016
D(lnII)	0.318	0.115	2.765	0.009
LnRP	0.095	0.311	0.306	0.761
D(lnRP)	0.100	0.326	0.306	0.761
C	−6.648	11.195	−0.524	0.421
Coint. Eq (−1)	−1.049	0.145	−7.190	0.000
Panel B: Model Criteria				
R-Squared	0.599	Adjusted R Square	0.474	
Durbin-Waston	1.811	Jarque-Bera	2.105 (0.349)	
Breusch-Godfrey	1.545(0.229)	Ramsey Reset	0.86 (0.80)	
Breusch-Pagan-Godfrey	0.972 (0.459)			
CUSUM	Stable	CUSUM Square	Stable	
Source: Author's Computations.				

Table 4.
Outcomes of long and short intervals of time.

Greenhouse Gas Emanations has a coefficient value of −0.617 and a highly significant probability value of 0.002 which indicated that cereal production will decline when greenhouse gas concentration starts increasing in the atmosphere. Findings are

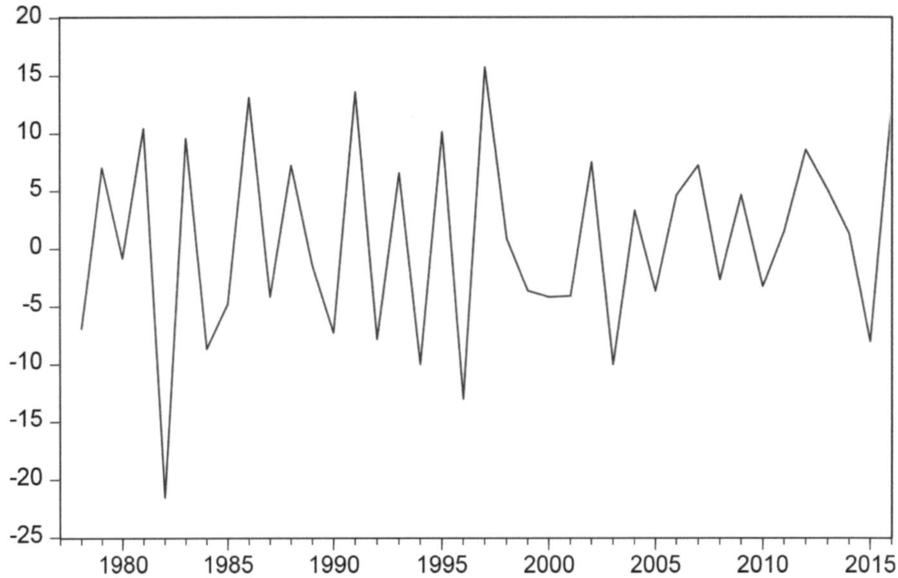

Figure 4.
Cointegration graph. Source: Author's computations.

in line with just CO_2 emission. Pickson et al. [74] employed ARDL on time series data (1990–2013) from cereal production in the case of China. Chandio et al. [79] also achieved the same result on the rice crop of Pakistan through ARDL from 1968 to 2014. Findings are contradicting [41, 42, 72, 79, 82]. While in the short run, it also has an adverse but insignificant liaison with cereal production such that 1kt will reduce 0.153 tonnes of cereal production this outcome is similar to [43, 46, 47, 74, 78].

Cultivated Area is known as a prime influencer for cereal production which has a coefficient value of 2.384 and a significant probability value of 0.017 and showed cereal production will increase in case of a rise in land under cereal production. Chandio et al. [41, 42, 44–47, 74, 78, 79, 83, 84] also provided similar result. The land under cereal production coefficient is 1.365 and significant which makes it possible that 1 per cent rise in the area will promote 1.365 per cent cereal production past studies; Chandio et al. [41, 42, 45, 47, 74, 79] supported it.

The input Index plays a pivotal character to boost and cope with the negative influence of climate change on cereal production over a long interval of time. The cereal input index coefficient is 0.303 with a probability value of 0.016 which exhibits that a 1 per cent increase in it will upsurge 0.303 cent cereal output. Proper practice of inputs helps to enhance soil fertility and nutrition and in addition is a major source of boosting production [41, 45, 46, 78, 79, 83]. In a short interval of time, cereal input index coefficient is 0.318 which means that a 1 per cent increase in it will increase 0.303 per cent cereal output and is similar to [41, 45, 83].

Rural Population is employed as a proxy of the rural labour force, indicating positive but insignificant linkages with a cereal production coefficient value is 0.095 which suggests that in the long interval of time, 1 per cent rise in it will increase 0.095 per cent cereal production. Results are matched by [42, 74, 77, 78] on labour force and [43] on rural population contradict by Warsame et al. [44]; and Sossou et al. [72]. Also, in a short interval of time, it has positive and insignificant such as a 1 per cent rise in it will ease -0.100 per cent cereal production [43].

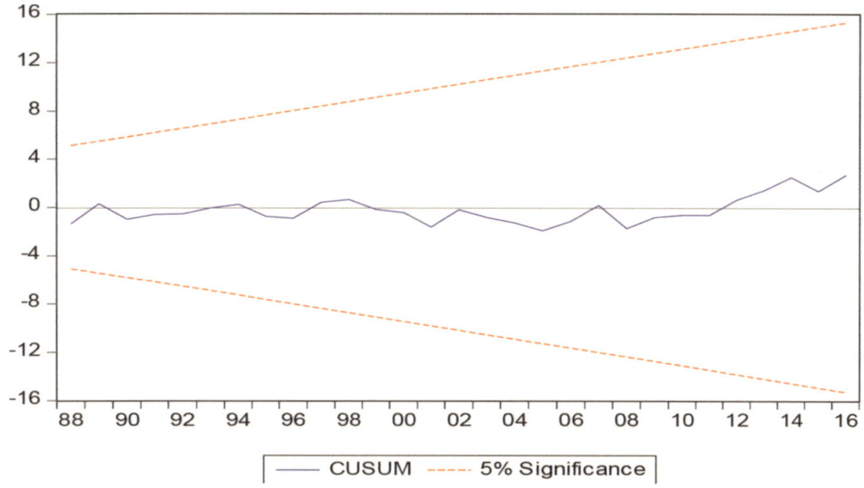

Figure 5.
Cumulative sum.

The Error Correction Coefficient is −1.049 and significant which infers a higher converging speed in the direction of equilibrium. This infers that the change in cereal production from short to long intervals of time is adjusted by almost 1% per year. The adjustment is highly significant because of the probability value. The input index plays a pivotal role to absorb any adverse shock in a short interval of time.

Furthermore, diagnostic inspection tests attempted to enhance the validity of analysis **Table 3**, and panel b displayed the outcome of it, which accessed that the cereal production model is normally specified, functional form and free from serial correlation, as well as heteroscedasticity. Mutually cooperated stability tests CUSUM[6] and CUSUMQ[7] plots specified that the cereal production model is stable and cannot be rejected, which are displayed in **Figures 5** and **6**. Furthermore, The pore over model is the goodness of fit for the reason that R^2:0.599 and Adj- R^2:0.474 suggest that very nearly 59% variations in cereal production are expounded by temperature index, precipitation index, greenhouse gases emanations, cultivated area, input index and rural population as well as a remaining error term.

5. Conclusion and policy relevance

Pakistan is known as a climate change persuaded hazard-prone country. Due to its geographical position, climate change manifests in the form of temperature rise, deplete glaciers, storms, heavy rainfall, an upsurge in sea level, typhoons, flash floods, glacial lake outbursts, unusual smog, droughts and landslides. The agriculture sector has got a crucial position in Pakistan's economy, and climate change has greatly influenced its production as compared to rest of the sectors. This article has drawn attention towards the vulnerability of climate change and the significance of adaptation and mitigation-based policies. Numerous models of climate change scenarios

[6] *cumulative sum*

[7] *cumulative sum of square*

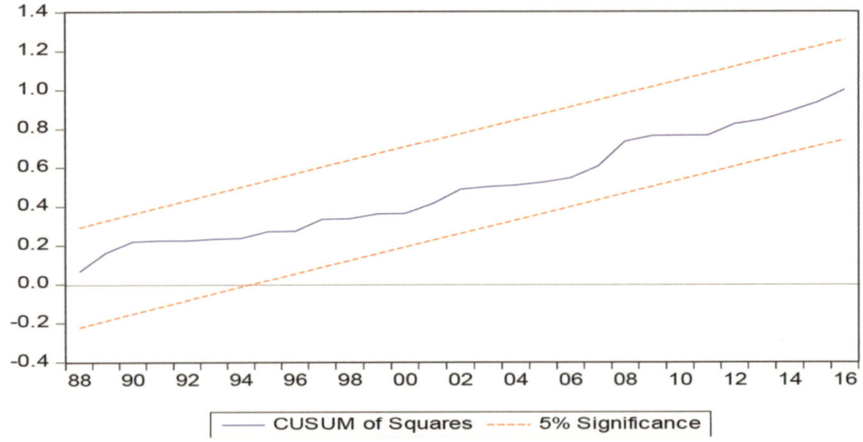

Figure 6.
Cumulative sum of square.

draw attention to the decline in cereals production during the current situation. The L-ARDL analysis hindered that temperature and greenhouse gas emanation have negative bonds with cereals production in the long run although greenhouse gas emanation is just in short intervals. Pakistan is standing in the row of low greenhouse gas emissions countries and, on the other hand, is drastically disturbed by it. This research suggests some adaptation and mitigation strategies through which we can tackle climate change:

- The government needs to introduce a national climate fund and insurance schemes for poor farmers. Rice crops should be cultivated through the latest cultivation way with lower methane emission and nitrous oxide from agricultural soil.

- World Trade Organisation plays a pivotal role to reduce barriers to transferring low-carbon and environment-friendly technologies.

- Decline dependency on fossil fuels for energy purposes and transfer from fossil fuels to renewable energy resources (solar energy and hydropower) and the need to construct large hydropower projects according to present and future requirements.

- To mitigate greenhouse gas emissions, the government should introduce sustainable development policies. It is time for the administrative system to take strong steps to promote the Clean Development Mechanism to reduce the carbon footprint of different sectors and employ a carbon pollution tax on those who emit more than a limited amount of carbon dioxide.

Conflict of interest

"The authors declare no conflict of interest."

Annexure 1

A.1 Brief History of Legislation and Plans of Climate Change in Pakistan

On 25th September 2015, United Nations introduced sustainable development goals with the 13th Sustainable Development Goal being "Climate Action". The past and present governments are acquainted with this issue and adopt strategic measures at national and international levels to abate climate change effects. Furthermore, climate change concerns are overcome with the help of endorsing legislation and generating strategies and policies instead of a healthy environment. The leading legislations and strategies to combat the climate change challenge are as under:

i. In 1983, Pakistan Environment Protection Ordinance was endorsed.

ii. During 1991–1993, National Environmental Quality Standards were prepared by the National Conservation Strategy adopted in 1993.

iii. In 1997, Pakistan's Environmental Policy derived from the Pakistan Environment Protection Act.

iv. During 2004–2005, Prime Minister Committee on Climate Change convenes National Environment Policy.

v. For the duration of 2008, the Planning Commission recognised the task force and Inter-Ministerial Committee introduced on climate change.

vi. Ministry of Disaster Management was renamed and converted into the Ministry of Climate Change

vii. In 2012, National Climate Change Policy introduced a plan to talk about climate change. And the National Disaster Risk Reduction Policy was also introduced

viii. Climate Change Policy Implemented framework for the duration of 2014 Framework.

ix. In 2015, the Division of Climate Change upgraded to the Ministry of Climate Change, Ministry Of Climate Change designed Forest Policy.

x. Green Pakistan Programme was designed for the plantation of 100 million plants all over the country.

xi. In the course of 2016, the government of Pakistan endorsed the Paris Climate Change Agreement in addition to the announced Climate Change Bill.

xii. Pakistan Climate Council was developed for the implementation of the Kyoto Protocol and the Paris Agreement.

xiii. During 2014–2018, the government took some positive steps to reduce carbon emissions, especially the adoption of Euro-II standard fuel, switching to RON

92 petrol from RON 87, banning the import of furnace oil for power generation and plans to CAP coal-fired power generation plants.

xiv. For the year 2019–2023, the Ten Billion Tree Tsunami Programme was launched which will plant 3.296 billion 1586.18 million planted in the whole of Pakistan.

Classification

Jel code: Q1, Q22, Q23, Q25, Q54.

Author details

Abdul Rehman Aslam* and Fatima Farooq
School of Economics, Bahauddin Zakariya University, Multan, Pakistan

*Address all correspondence to: abdulrehmanaslam6811@gmail.com

References

[1] Ullah W, Takaaki N. Climate change vulnerability of Pakistan towards natural disasters: A review. International Journal of Environmental Protection and Policy. 2016;**4**(5):1-7

[2] Government of Pakistan. Pakistan Economic Survey 2017–18. Pakistan: Ministry of Finance; 2018

[3] Stern N. The Economics of Climate Change – The Stern Review. U.K.: Cambridge University Press; 2006

[4] Wallace B. Climatic change: Are we on the brink of a pronounced global warming? Science. 1975;**189**:460-463

[5] NRC. Advancing the Science of Climate Change. National Research Council: The National Academies Press, Washington, DC, USA; 2010

[6] Rahmstorf S. A semi-empirical approach to projecting future sea-level rise. Science. 2007;**315**(5810):368-370

[7] Stocker FT, Dahe Q, Plattner G. Climate change 2013: The physical science basis. In: Working Group I Contribution to the Fifth Assessment Report of the Intergovernmental Panel on Climate Change Summary for Policymakers. 2013

[8] Pörtner HO, Roberts DC, Adams H, Adler C, Aldunce P, Ali E, et al. Climate Change 2022: Impacts, Adaptation and Vulnerability. Netherland: IPCC Sixth Assessment Report; 2022

[9] Ahmed M, Suphachalasai S. Assessing the Costs of Climate Change and Adaptation in South Asia. Mandaluyong City, Metro Manila, Philippines: Asian Development Bank; 2014

[10] Chaudhry QUZ. Climate Change Profile of Pakistan. Mandaluyong City, Metro Manila, Philippines: Asian Development Bank; 2017

[11] Farooqi AB, Khan AH, Mir H. Climate change perspective in Pakistan. Pakistan Journal of Meteorology. 2005; **12**(3):11-21

[12] Hoerling M, Kumar A. The perfect ocean for drought. Science. 2003;**299** (5607):691-694

[13] Larsen O, Oliver J, Casiles Lanuza E. Developing a Disaster Risk Insurance Framework for Vulnerable Communities in Pakistan: Pakistan Disaster Risk Profile. Report No. 16. Bonn: United Nations University Institute for Environment and Human Security (UNU-EHS); 2014

[14] Government of Pakistan. National Consultative Workshop on Pakistan's Intended Nationally Determined Contribution (Pak-INDCs). Government of Pakistan; 2015

[15] Government of Pakistan. Various Annual Report. Pakistan: National Disaster Management Authority; 2013-2021

[16] Bhutta ZA, Bhutta SZ, Raza S, Sheikh AT. Addressing the human costs and consequences of the Pakistan flood disaster. The Lancet. 2022;**400**(10360): 1287-1289

[17] Shahzad A. U.N. Chief Calls for 'Massive' Help as Pakistan Puts Flood Losses at $30 Billion. Pakistan: Reuters; 2022

[18] Zahid M, Rasul G. Changing trends of thermal extremes in Pakistan. Climatic Change. 2012;**113**(3-4):883-896

[19] Khan N, Shahid S, Ismail T, Ahmed K, Nawaz N. Trends in heat wave related

indices in Pakistan. Stochastic Environmental Research and Risk Assessment. 2019;**33**(1): 287-302

[20] Khan TMA, Rabbani MM. Sea Level Monitoring and Study of Sea Level Variations along Pakistan Coast: A Component of Integrated Coastal Zone Management. Karachi, Pakistan: National Institute of Oceanography; 2000

[21] Betz J, Scholvin S. Energy Policy in the Emerging Economies: Climate Change Mitigation under the Constraints of Path Dependence. London, United Kingdom: Rowman & Littlefield Publishing Group, Inc.; 2019

[22] Kakar A, Hayat MT, Abbasi AM, Pervez A, Mahmood Q, Farooq U, et al. Risk assessment of heavy metals in selected marine fish species of Gadani shipbreaking area and Pakistan. Animals. 2020;**10**(10):1738

[23] Government of Pakistan. Pakistan Economic Survey 2020–21. Pakistan: Ministry of Finance; 2022

[24] Dehlavi A, Groom B, Gorst A. Climate Change Adaptation in the Indus Eco Region: A Micro-Econometric Study of the Determinants, Impacts and Cost Effectiveness of Adaptation Strategies. Pakistan, Islamabad: World Wide Fund for Nature (WWF); 2015

[25] Athar H. Water supply and effective rainfall impacts on major crops across irrigated areas of Punjab. Pakistan. Theoretical and Applied Climatology. 2020;**142**(3):1097-1116

[26] Ahmad S, Islam M, Mirza SN. Rangeland degradation and management approaches in Balochistan, Pakistan. Pakistan Journal of Botany. 2012;**44**: 127-136

[27] Mansoor M, Jamil M, Anwar F, Awan AA, Muhammad S. Review a review on rangeland Management in Pakistan. Bottlenecks and Recommendations. Biological Sciences-PJSIR. 2018;**61**(2):115-120

[28] Mir KA, Ijaz M. Greenhouse Gas Emission Inventory of Pakistan for the Year 2011–2012. Islamabad, Pakistan: Global Change Impact Study Centre; 2015 GCISC Interim Research Report RR-19 (Interim)

[29] Government of Pakistan, Pakistan Meteorological Department (2015b). High Resolution Climate Scenarios. Available from: http://www.pmd.gov.pk/rnd/rnd web/rnd_new/climchange_ar5.php

[30] Malik SM, Awan H, Khan NU. A study of the effects of climate change on human health in Pakistan-Evidence based policy advocacy. 2010. Available from: Sightsavers website: http://pk.sigh tsavers.Org/in_depth/policy_and_resea rch/15837_climatechange.pdf

[31] Khalid B, Ghaffar A. Dengue transmission based on urban environmental gradients in different cities of Pakistan. International Journal of Biometeorology. 2015;**59**(3):267-283

[32] Trisos CH, Merow C, Pigot AL. The projected timing of abrupt ecological disruption from climate change. Nature. 2020;**580**(7804):496-501

[33] Government of Pakistan. Ministry of Planning, Development and Reforms (2010). Islamabad: Task Force Final Report on Climate Change; 2010

[34] Verge XP, De Kimpe C. Agricultural production, greenhouse gas emissions and mitigation potential. Agricultural and Forest Meteorology. 2007;**142**(2-4): 255-269

[35] Gowdy J. Climate change and economic development: A pragmatic approach. The Pakistan Development Review. 2007;**46**(4):337-350

[36] Ahmed M, Schmitz M. Economic assessment of the impact of climate change on the agriculture of Pakistan. Business and Economic Horizons. 2011; **4**:1-12

[37] Siddiqui R, Samad G, Nasir M, Jalil HH. The impact of climate change on major agricultural crops: Evidence from Punjab, Pakistan. The Pakistan Development Review. 2012;**51**(4): 261-274

[38] Janjua PZ, Samad G. Climate change and wheat production in Pakistan: An autoregressive distributed lag approach. NJAS-Wageningen Journal of Life Sciences. 2014;**68**:13-19

[39] Zhai S, Song G, Qin Y, Ye X. Modeling the impacts of climate change and technical progress on the wheat yield in inland China: An autoregressive distributed lag approach. PLoS One. 2017;**12**(9):0184474

[40] Simionescu M, Bilan Y, Gędek S. The effects of greenhouse gas emissions on cereal production in the European Union. Sustainability. 2019;**11**(12):3433

[41] Chandio AA, Jiang Y, Rehman A, Rauf A. Short and long-run impacts of climate change on agriculture: An empirical evidence from China. International Journal of Climate Change Strategies and Management. 2020a;**12** (2):201-221

[42] Ahsan F, Chandio AA, Fang W. Climate change impacts on cereal crops production in Pakistan: Evidence from cointegration analysis. International Journal of Climate Change Strategies and Management. 2020;**12**(2):257-269

[43] Chandio AA, Ozturk I, Akram W, Ahmad F, Mirani AA. Empirical analysis of climate change factors affecting cereal yield: Evidence from Turkey. Environmental Science and Pollution Research. 2020c;**27**(11):11944-11957

[44] Warsame AA, Sheik-Ali IA, Ali AO, Sarkodie SA. Climate change and crop production nexus in Somalia: An empirical evidence from ARDL technique. Environmental Science and Pollution Research. 2021;**28**(16):19838-19850

[45] Abbas S. Climate change and major crop production: Evidence from Pakistan. Environmental Science and Pollution Research (Switzerland: Springer Nature). 2022;**29**(4):5406-5414

[46] Baig IA, Chandio AA, Ozturk I, Kumar P, Khan ZA, Salam M. Assessing the long-and short-run asymmetrical effects of climate change on rice production: Empirical evidence from India. Environmental Science and Pollution Research. 2022;**29**:1-22

[47] Chandio AA, Jiang Y, Fatima T, Ahmad F, Ahmad M, Li J. Assessing the impacts of climate change on cereal production in Bangladesh: Evidence from ARDL modeling approach. International Journal of Climate Change Strategies and Management. 2022;**14**(2): 125-147

[48] Adger WN. Vulnerability. Global Environmental Change. 2006;**16**(3): 268-281. DOI: 10.1016/j.g

[49] Adger WN, Huq S, Brown K, Conway D, Hulme M. Adaptation to climate change in the developing world. Progress in Development Studies. 2003;**3** (3):179-195

[50] Brooks N. Vulnerability, risk and adaptation: A conceptual framework.

Tyndall Centre for Climate Change Research Working Paper. 2003;**38**:1-16

[51] Füssel HM. Vulnerability: A generally applicable conceptual framework for climate change research. Global Environmental Change. 2007;**17** (2):155-167

[52] Füssel HM, Klein RJ. Climate change vulnerability assessments: An evolution of conceptual thinking. Climatic Change. 2006;**75**(3):301-329

[53] Helmer M, Hilhorst D. Natural disasters and climate change. Disasters. 2006;**30**(1):1-4

[54] Lindner M, Maroschek M, Netherer S, Kremer A, Barbati A, et al. Climate change impacts, adaptive capacity, and vulnerability of European forest ecosystems. Forest Ecology and Management. 2010;**259**(4):698-709

[55] McCarthy JJ. Climate Change 2001: Impacts, Adaptation, and Vulnerability: Contribution of Working Group II to the Third Assessment Report of the Intergovernmental Panel on Climate Change. UK: Cambridge University Press; 2001

[56] Mirza MMQ. Climate change and extreme weather events: Can developing countries adapt? Climate Policy. 2003;**3** (3):233-248

[57] O'brien K, Eriksen S, Nygaard LP, Schjolden A. Why different interpretations of vulnerability matter in climate change discourses. Climate Policy. 2007;**7**(1):73-88

[58] Smit B, Pilifosova O. From adaptation to adaptive capacity and vulnerability reduction. In: Climate Change, Adaptive Capacity and Development. London: Imperial College Press; 2003. pp. 9-28. ISBN 1-86094-373-X

[59] Smith JB, Huq S, Klein RJ. Climate Change, Adaptive Capacity and Development. London: Imperial College Press; 2003

[60] Schaeffer R, Szklo AS, de Lucena AFP, Borba BSMC, Nogueira LPP, Fleming FP, et al. Energy sector vulnerability to climate change: A review. Energy. 2012;**38**(1):1-12

[61] Thomas K, Hardy RD, Lazrus H, Mendez M, Orlove B, Rivera-Collazo I, et al. Explaining differential vulnerability to climate change: A social science review. Wiley Interdisciplinary Reviews: Climate Change. 2019;**10**(2): e565

[62] Thornton PK, Ericksen PJ, Herrero M, Challinor AJ. Climate variability and vulnerability to climate change: A review. Global Change Biology. 2014;**20** (11):3313-3328

[63] Cutter SL, Boruff BJ, Shirley WL. Social vulnerability to environmental hazards. Social Science Quarterly. 2003; **84**(2):242-261

[64] Smit B, Wandel J. Adaptation, adaptive capacity and vulnerability. Global Environmental Change. 2006;**16** (3):282-292

[65] Blaikie P, Cannon T, Davis I, Wisner B. At Risk: Natural Hazards, People's Vulnerability and Disasters. London: Routledge; 2014

[66] Adams RM, McCarl BA, Dudek DJ. Implications of global climate change for western agriculture. Western Journal of Agricultural Economics. 1988;**13**(2): 348-356

[67] Callaway JM. An Analysis of Methods and Models for Assessing the Direct and Indirect Economic Impacts of CO2 Induced Environmental Changes in

the Agricultural Sector of US Economy. Washington: Pacific Northwest Laboratory; 1982

[68] Decker WL, Jones VK. The Impact of Climate Change from Increased CO2 on American Agriculture. Washington, DC: US Department of Energy; 1986

[69] Afzal M, Ahmed G, Javaid M. Emperical Asessment of climate change on agriculture crops: Panel data analysis in Pakistan. International Journal of Food and Agricultural Economics. 2017;**5** (2):59-68

[70] Zhengfei G. Econometric Analysis of Agricultural Production: New Primal Perspectives. Netherlands: Ponsen & Looijen BV; 2005

[71] Chambers RG. Applied Production Analysis: A Dual Approach. New York, USA: Cambridge University Press; 1988

[72] Sossou S, Igue CB, Diallo M. Impact of climate change on cereal yield and production in the Sahel: Case of Burkina Faso. Asian Journal of Agricultural Extension, Economics & Sociology. 2019;**37**(4):1-11

[73] Lu S, Bai X, Li W, Wang N. Impacts of climate change on water resources and grain production. Technological Forecasting and Social Change. 2019;**143**: 76-84

[74] Pickson RB, He G, Ntiamoah EB, Li C. Cereal production in the presence of climate change in China. Environmental Science and Pollution Research. 2020;**27**: 45802-45813. DOI: 10.1007/s11356-020-10430-x

[75] Pesaran M, Shin Y. An autoregressive distributed-lag modelling approach to cointegration analysis. Econometric Society Monographs. 1998; **31**:371-413

[76] Pesaran M, Shin Y, Smith R. Bounds testing approaches to the analysis of level relationships. Journal of Applied Econometrics. 2001;**16**(3):289-326

[77] Chandio AA, Jiang Y, Akram W, Adeel S, Irfan M, Jan I. Addressing the effect of climate change in the framework of financial and technological development on cereal production in Pakistan. Journal of Cleaner Production. 2021b;**288**:125637

[78] Chandio AA, Gokmenoglu KK, Ahmad M, Jiang Y. Towards sustainable rice production in Asia: The role of climatic factors. Earth Systems and Environment. 2021;**6**:1-14

[79] Chandio AA, Magsi H, Ozturk I. Examining the effects of climate change on rice production: Case study of Pakistan. Environmental Science and Pollution Research. 2020b;**27**(8):7812-7822

[80] Attiaoui I, Boufateh T. Impacts of climate change on cereal farming in Tunisia: A panel ARDL–PMG approach. Environmental Science and Pollution Research. 2019;**26**(13):13334-13345

[81] Jan I, Ashfaq M, Chandio AA. Impacts of climate change on yield of cereal crops in northern climatic region of Pakistan. Environmental Science and Pollution Research. 2021;**28**(42):60235-60245

[82] Chandio AA, Jiang Y, Joyo MA, Pickson RB. Research on factors influencing the grain production in Pakistan: An ARDL approach. European Online Journal of Natural and Social Sciences. 2018;**7**(3):538

[83] Chandio AA, Jiang Y, Joyo MA. Impact of area under cultivation, water availability, credit disbursement, and fertilizer off-take on wheat production

in Pakistan. Journal of Applied Environmental and Biological Science. 2016;**6**(10):10-18

[84] Chandio AA, Jiang Y, Rehman A. Using the ARDL-ECM approach to investigate the nexus between support price and wheat production. Journal of Asian Business and Economic Studies. 2019;**26**(1):139-152

Chapter 12

Knowledge Coproduction for Transformative Climate Adaptation: Building Robust Strategies

Yosune Miquelajauregui and Adela Madariaga-Fregoso

Abstract

Adaptation is a process of adjustment to actual or expected climate and its effects in order to moderate harm or exploit beneficial opportunities. Most adaptation options are scalable and applicable but may result in inequitable tradeoffs stemming from maladaptation. Thus, climate adaptation and maladaptation are inseparable and are equally likely. Adaptation has been commonly envisioned as coping mechanisms or incremental adjustments from existing strategies. However, both coping and incremental adaptations have failed in explicitly address the underlying drivers of systemic inequalities. Enabling and catalyzing conditions for transformative adaptation, both locally and regionally (i.e. strengthening collaborative governance, building capacities, promoting iterative multi-stakeholder engagement), is, therefore, crucial in building robust climate change adaptations under deep uncertainty. However, the lack of approaches entailing decision analytics, stakeholder engagement/deliberation, and interactive modeling and evaluation may hinder transformative adaptation success. Combining robust decision-making approaches with collaborative research and co-production processes can be constructive in illuminating the decision-rule systems that undergird current adaptation decision-making. This chapter offers some insights into how knowledge coproduction can be used to inform robust climate adaptation strategies under contexts of deep uncertainty while facilitating transformative system change.

Keywords: climate change adaptation, knowledge coproduction, transformations, robust decision-making, sustainability

1. Introduction

Climate change is a multicausal, technically complex, controversial, and highly uncertain problem [1]. The ability of coupled human-earth systems to adapt to the direct and indirect impacts of climate change is, therefore, critical in order to achieve sustainability [2–4]. Climate change adaptation became a popular concept among scholars after the United Nations climate change convention in the 1990s. From that point, climate change adaptation has been implemented as coping mechanisms that tend to focus only on proximate causes, as well as incremental adjustments of existing institutional, financial, and technological adaptation strategies [5–7]. As climate change intensifies, fundamental shifts in existing resource systems, policies, power dynamics, and stakeholders'interests and mindsets will be required if we are to keep the average rise in temperature below 2°C [8, 9]. However, both coping and incremental adaptation strategies are not sufficient to promote these long-lasting system transformations [9–11]. Transformative climate adaptation, on the other hand, has the potential to respond to the magnitude of cascading climate risks by facilitating radical shifts in coupled human-earth systems.

In coupled human-earth systems, characterized by interlocked multisector interactions and feedbacks (e.g. environmental, socio-economic, technological, governance, and institutional), climate change adaptation planning is further complicated by high degrees of uncertainties [12]. Uncertainty emerges from the limited and contested knowledge among stakeholders regarding (i) the appropriate models to describe the key drivers of the system (e.g. population growth, urban sprawl, and water demand), (ii) the probability distributions about key variables and parameters, and (iii) the relative importance of alternative outcomes (e.g. trade-offs among goals) [12, 13]. According to Ref. [13], uncertainty also arises from human actions, which are taken in response to unpredictable situations over time. In order to manage uncertainty in an efficient way, adaptation planning should be able to confront and navigate alternative adaptation strategies in order to choose robust ones that perform well over a wide range of plausible futures.

Adaptation planning cannot be successfully addressed with traditional linear analytical approaches. In this regard, the field of Decision Making Under Deep Uncertainty (DMDU) has emerged as a promising framework that supports and informs climate change adaptation planning under uncertainty [7, 12, 13]. DMDU includes a set of approaches including *Robust Decision-Making* (RDM), *Dynamic Adaptive Planning* (DAP), *Dynamic Adaptive Policy Pathways* (DAPP), and *Info-Gap Decision Theory* (IG). These approaches accentuate the transition from classical "predict then act" risk management to exploratory modeling. In particular, RDM explicitly follows a learning process called deliberation with analysis that supports decision-makers and stakeholders to iteratively and collaboratively frame the adaptation problem, specify performance metrics and modeling methods, design the experimental framework, evaluate the performance of strategies across multiple futures, and choose or modify robust adaptation strategies [12, 14].

However, climate change adaptation planning is generally built on divergent stakeholder interests and disparate problem framings, meaning that planners do not always agree on common problem definitions and plausible pathways to adaptation [1, 14, 15]. Moreover, adaptation planning is also embedded within political, social, and institutional contexts that shape how networks of actors interact through formal and informal relationships, rules, and norms [15]. In this perspective, collaborative research and coproduction processes can be constructive in illuminating the decision-rule systems that undergird current stakeholder decision-making and revealing how they are, or are not, functioning to deliver desired results, helping stakeholders interrogate what their preferences are and how those preferences can or cannot be met under a wide variety of conditions [13]. Knowledge coproduction has been acknowledged as an action-oriented practice that enables consensus, coordination, and transparency among stakeholders, thus enhancing policy-relevant climate knowledge [16–18].

This chapter exposes the readers with a synthesis of the state-of-the-art theory and practice associated with climate change adaptation planning under deep uncertainty. The text consists of three subsections. The first subsection presents a review of the different adaptation strategies and their scope in addressing key drivers of systemic inequality. The second subsection presents an overview of robust decision-making (RDM) approaches illustrated through a hypothetical case study for transformative adaptation planning. The last subsection presents some insights into how knowledge coproduction can be used to inform robust climate adaptation strategies under contexts of deep uncertainty.

2. Knowledge coproduction for climate change transformative adaptation: building robust strategies

2.1 Coping, incremental, and transformative climate change adaptations

Adaptation to climate change is an ongoing process by which coupled human-earth systems adjust in response to the observed and expected climatic stimuli in order to better manage the risks posed by climate change [2, 3]. Adapting to climate change involves cascading decisions that transverse multiple sectoral – infrastructure (i.e freshwater, energy, food, and health), governing institutions (public and private), and regulatory agencies – and geographical boundaries (i.e. local, regional, and global), underscoring the complexity of this global phenomenon [19–21]. Adaptation takes place across multiple temporal and spatial scales and usually entails vulnerability and risk assessments, identification of strategies, planning, monitoring, evaluation, and review [4]. In the climate-change literature, adaptation has been commonly envisioned as a mechanism to cope with risks (e.g. borrowing money to repair houses) and as increments of existing adaptation strategies (e.g. building higher dams or resistant buildings) aimed at accommodating change [10, 21]. However, both coping and incremental adaptation strategies have generally failed to directly address the underlying drivers of systemic inequalities in climate change impacts, that is, to deliberately and fundamentally change systems to achieve equitable distribution of adaptation outcomes (**Figure 1**) [10, 11, 21].

Climate change adaptation planning takes place in a context of multiple uncertainties including epistemic (i.e. imperfection of knowledge), normative (i.e. impossibility of knowing the evolution of ethical values), political-induced (i.e. deliberative ignorance of public agencies), knightian (i.e. impossibility ok knowing all the information), and deep uncertainty (i.e. disagreement about the adequacy of models) [22]. Moreover, as a process, climate change adaptation is highly controversial since it usually entails the participation of multiple stakeholders with asymmetric power and competing knowledge [1, 15, 22]. Thus, planning for transformative climate change adaptation must explicitly address the historical power struggles and imbalances, the goals and mindsets of powerful actors, and the structure and rules-in-use that shape system dynamics [1, 10, 11]. However, transformative adaptation planning often encounters multiple barriers that can dampen efforts to create long-term climate-robust adaptation strategies [14].

For example, Ref. [3] and Ref. [14] identified a set of external and internal barriers to transformative adaptation development and implementation. External barriers include, for instance, the uncertainties related to climate change projections, future distribution of extreme meteorological events, expected risks, and vulnerability

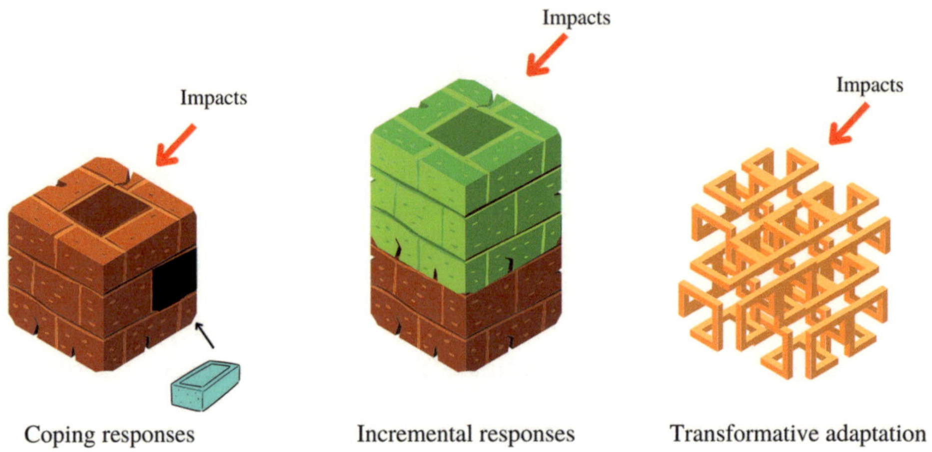

Figure 1.
Coping, incremental, and transformative climate change adaptation.

outcomes, as well as potential adaptation costs and benefits of climate change policy instruments. Internal barriers relate to the interaction and feedback across the existing institutional arrangements and governance structures underpinning system's dynamics and which operate within specific political, cultural, and social contexts. Moreover, a number of internal barriers to effective decision-making have also been highlighted by Ref. [1] and Ref. [16] including, for instance, analytical difficulties in explicitly incorporating multiple stakeholders' values, interests, and attitudes into socially-accepted and climate-resilient adaptation plans.

Transformative adaptation entails challenging the status-quo of the current system by fundamentally changing the material (e.g. policies and practices), procedural/relational (e.g. power dynamics), and conceptual/cognitive (e.g. values and preferences) dimensions of human-earth systems (**Figure 2**) [8, 11]. These changes include, for instance, modifying economic paradigms and development patterns, decolonizing knowledge systems, reforming governance institutions, and transforming the relationships and power dynamics among actors, as well as enabling individual and collective empowerment through learning and knowledge coproduction [11, 15, 16].

2.2 Robust decision-making (RDM) for transformative climate adaptation

Complex and uncertain sustainability issues, such as the climate change crisis, require the adoption of robust decision-making (RDM) approaches that help decision makers identify adaptation strategies that perform well over a wide range of uncertain futures given socio-economic, environmental, political, and technological future trends [12, 13]. Following [13], "RDM explicitly follows a learning process called deliberation with analysis that promotes learning and consensus-building among stakeholders." In this perspective, deliberation with analysis entails the coproduction of policy-relevant and legitimate knowledge.

In the context of climate change, the RDM deliberative process starts by eliciting stakeholders' priorities, preferences, and assumptions underpinning adaptation planning [9, 15]. As noted by Ref. [15], transparency concerning the norms and procedures for deliberation is critical to maximize consensus and minimize conflicts among the stakeholders. Deliberation with analysis requires participants

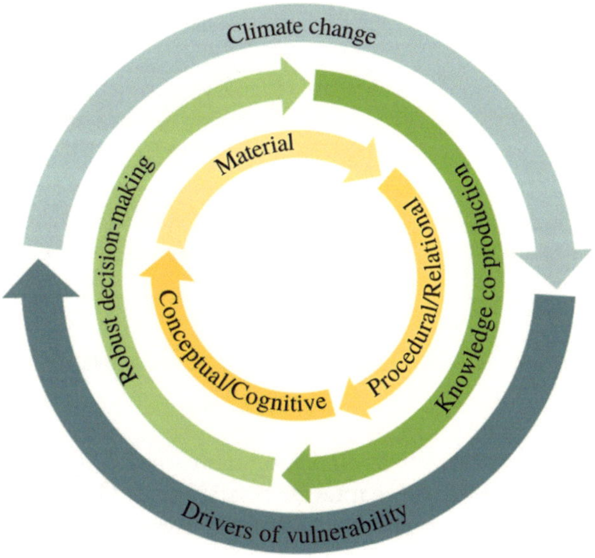

Figure 2.
Transformative climate adaptation entails radical changes in the material (e.g. policies and practices), procedural/relational (e.g. power dynamics), and conceptual/cognitive (e.g. values and preferences) dimensions of human-earth systems through knowledge coproduction under conditions of deep uncertainty.

to actively engage with sharing their knowledge in order to collectively frame the decision problem, specify performance metrics and modeling methods, design the experimental framework, evaluate performance of strategies across futures, and choose or modify robust strategies [7, 12, 13]. **Table 1** presents a hypothetical case study for climate change adaptation planning. Through an RDM deliberative process, stakeholders identify relevant sources of uncertainties, system parameters, variables and relationships, analytical methods, alternative adaptation strategies, and performance metrics regarding an adaptation planning problem. This information is organized in a four-quadrant matrix called XLRM as follows: (1) exogenous uncertainties (X) including, for example, temperature and precipitation projections under multiple global climate models and radiative forcing scenarios, urban sprawl scenarios, socio-economic future trends, and changes in risk perceptions; (2) alternate adaptation strategies (L), which are driven by stakeholders; (3) causal relationships (R) to simulate interactions and dynamics of human-earth systems; and (4) performance metrics (M) to assess strategy performance and characterize robustness.

However, coordinating and implementing RDM deliberative processes may be challenging due to divergent and competing stakeholders` climate risk perceptions and attitudes [15]. Moreover, mainstreaming community, indigenous and local climate change risk perceptions and attitudes into climate change adaptation planning has generally been uncommon in RDM processes [23–25]. Enabling tools such as computer platforms, computer-based scenarios, and visualization techniques (i.e. boundary objects [16]) offer some potential to facilitate RDM deliberative processes. These boundary objects can be used to iteratively integrate stakeholders' values, interests, and knowledge examine the main processes and uncertainties affecting human-earth system's dynamics, and collectively identify potentially vulnerable zones to climate change impacts [12, 13]. Interaction with these decision support tools

Uncertainties (X)	Levers (L)
Climate change projections Urban sprawl GDP /socioeconomic trends Risk perceptions	Current adaptation strategy Strategy including tech. Inn. Strategy including carbon incentives Strategy including tech. Inn. & agency
Relationships (R)	Metrics (M)
Spatially-explicit simulation model (MEGADAPT) to evaluate urban socio- hydrological vulnerability	Reduction in socio-hydrological vulnerability Increase in adaptive capacities cost

Table 1.
XLRM matrix showing the main elements of a hypothetical RDM deliberative process.

helps stakeholders to formalize their value judgments regarding risk thresholds and uneven outcomes across alternate adaptation strategies [1, 15, 26].

As an "agree-on-decision" approach, RDM next uses simulation models (R) to evaluate selected adaptation strategies (L) over a wide range of uncertain futures (X) (**Table 1**). This step usually generates large databases of simulation model results [12, 13]. RDM advances the use of multi-objective optimization algorithms to trace out a range of potentially robust solutions [12–14]. Analysts and decision makers then apply visualization and data mining techniques including Patient Rule Induction Method (PRIM) and Classification and Regression Tree (CART) algorithms on these large databases to explore and characterize uncertain factors that define vulnerabilities. The vulnerability analysis is an iterative and interactive exercise that allows stakeholders to better understand the system's conditions and uncertainties under which systemic failures in adaptation can take place [13, 27]. The goal of these analyses is to establish a basic plan that dynamically adapts to signposts over time [7, 28].

As a result of the vulnerability analysis, new adaptation strategies emerge and are reexamined. Alternative strategies can be crafted using expert knowledge, thus conveying a collective strategic vision of a desired future state and the ways to get there [28, 29]. As stated by Ref. [28], policy-making becomes then an essential component of the storyline. Nevertheless, these top-down strategies are usually constrained in their ability to represent disruptive social and technological innovations and imaginaries, meaning they are relatively unresponsive to the underlying drivers of systemic inequalities [29]. Knowledge coproduction processes have the potential to inform robust transformative climate adaptations by engaging with diverse knowledge systems in order to ground alternative adaptation strategies in local realities, perspectives, and visions [15, 29].

2.3 Knowledge co-production for climate change adaptation planning

In the last decades, academic scholars have recognized the urgency to transform traditional science-practice relationships into action-oriented collaborative research in order to effectively address the most pressing challenges of society [17, 18]. Knowledge coproduction is part of this evolving set of action-oriented approaches defined by Ref. [30] as an "iterative, interactive and collaborative process involving diverse types of expertise, knowledge, and actors to produce context-specific knowledge." This definition underscores the normative aspirations underpinning collaborative scientific practices aimed at transcending the narrowness of disciplinary worldviews through the inclusion of diverse societal actors' perspectives, discourses, expertise, beliefs, and interests to solve complex and uncertain problems [1, 17, 30–34].

Knowledge coproduction is a context-based, pluralistic, goal-oriented, and iterative process that bridges knowledge to action [9, 18, 33]. Thus, coproduction entails multi-stakeholder (i.e. academic and nonacademic) engagement and participation to generate impact-driven information, which is sufficiently credible, legitimate, and salient. In accordance with Ref. [16], credibility concerns the technical adequacy of information, legitimacy refers to the perception that the production of information has been respectful for stakeholders' interests and needs, and salience refers to the relevance of the information. By directly connecting science, policy, and action, coproduction not only generates salient, credible, and legitimate knowledge to define adaptive interventions but also capacities (e.g. technical, analytical procedural, and evaluative), actor network partnerships, and inter-institutional organization fundamentally required for transformative climate change adaptations [10, 11, 21, 22].

Coproduction has been widely incorporated in the fields of health, education, development, and environmental planning [32]. Though the rate of use of coproduction processes in decision-making remains below expected needs, increased public participation in climate change adaptation planning has been recently reported in peer-reviewed scholarly publications [32–35]. In a comprehensive literature review of more than a hundred scientific publications on climate change, Ref. [35] found no common notion of coproduction but rather a broader collection of conceptual lenses from which coproduction is conceived and implemented. In this perspective, the lenses shed light on the diversity of coproduction goals, theories, practices, capacities, and outcomes providing key insights for policy making [33–35]. Despite differences in how the coproduction lenses interweave knowledge and action, the concept of boundary work was shown to serve as a common tool to both account for conflicting interests between political, social, and environmental externalities and to systematically reflect on the normative and participatory dimensions of the decision-making process (**Table 2**) [26, 30, 33, 35].

Climate change adaptation planning is generally built on uncertainty, divergent stakeholder interests, disparate problem framings, and dynamic socio-environmental interactions [1, 15, 19, 20]. Given the "wicked" nature of climate change adaptation planning, deterministic approaches are highly inappropriate since they can give rise to maladaptation and increase climate change vulnerabilities [27]. Vulnerability refers here to the susceptibility to being harmed by climate change, and it dynamically differs within communities and across regions and countries [36]. Research suggests that vulnerability stems from historic structural inequalities, power dynamics and legacies of past interventions, uneven resource distribution (e.g. water, housing, land), centralization of political power, systemic racism, and preexisting social and cultural norms that reinforce unsustainable paradigms [1, 11, 15, 20]. Consequently, advancing climate change transformative adaptation planning demands effectively responding to the magnitude of climate hazards while addressing the drivers of inequality in order to achieve successful outcomes under a wide range of uncertainties and different operational planning periods [7, 11, 28, 37].

International initiatives such as the *World Bank Climate Change Action Plan* and the *NOAA Regional Integrated Science Assessment* (RISA) have the potential to foster transformative adaptation planning through interdisciplinary research and engagement [34]. The former represents a global effort to support participation in key partnerships and forums aimed at improving climate change adaptation and resilience

Lense	Description	Application
Iterative interaction	Relative to the usability of climate information products in a decision-making context	Transform climate science into value-added "climate services"
Extended science	Looks at ways of doing science differently by including the knowledge and values of nonscientists	Democratizing practices of transdisciplinary science to generate robust climate knowledge
Public services	Joint production of public goods and services by government agencies and citizens	Institutional economics and multilevel governance
Institutional	Looks at how the processes of knowledge coproduction build adaptive capacities within governance institutions	Political ecology and environmental science
Social learning	Looks at how coproduction facilitates social learning about climate issues	Organizational studies, policy research, and management theory
Empowerment	Looks at the ways coproduction recognizes and empowers traditional knowledge systems	Anthropology, philosophy of science, and resource management

Table 2.
Normative lenses of coproduction, descriptions, and applications according to Ref. [35].

while addressing, to some, the root causes of vulnerability. The latter, on the other hand, represents a regionally-focused initiative aimed at building sustained partnerships to support equitable and collaborative adaptation to climate change risks [34, 37]. Other examples pertaining the Global South include the *binational laboratory on sustainability, vulnerability, and adaptation to Climate Change* (SVACC), a university-based node of collaboration among the US, Mexico, Central America, and the Caribbean aimed at strengthening regional technical and institutional capacities for effective climate change collaborative governance [26]. Despite institutional efforts to drive climate actions by actively and interactively engaging with stakeholders, the climate change research community has not yet achieved a shared conceptual and methodological decision support framework for collaboratively identifying sources of uncertainty, assessing (weight and appraise) risks and vulnerabilities, navigating and prioritizing risk–benefit adaptation trade-offs, choosing among adaptation strategies, and finally evaluating outcomes throughout the decision-process [3, 15].

3. Conclusions

Knowledge of coproduction processes is required if we are to navigate climate change uncertainty and support evidence-based adaptation policy-making. Large-scale systemic thinking at the material (e.g. policies and practices), procedural/relational (e.g. power dynamics), and conceptual/cognitive (e.g. values and preferences) dimensions of human-earth systems is increasingly promoted as a means of enhancing transformative climate change adaptation. Robust decision-making approaches are grounded on a learning process called deliberation with analysis. Deliberation with analysis entails the coproduction of knowledge to support decision-makers and stakeholders to frame the decision problem, specify performance metrics and modeling methods, design the experimental framework, evaluate performance of strategies across futures, and choose or modify robust strategies. Yet, competing knowledge and perceived injustice can dampen efforts to bring together academic and nonacademic actors in the process. Knowledge coproduction can help to navigate these challenges by making tangible and tractable issues of equity and justice in climate change adaptation planning.

Acknowledgements

This work was supported by the Universidad Nacional Autónoma de México (UNAM) [PAPIIT- Proyecto IN223321 Modelación de la resiliencia de servicios ecosistémicos en el Suelo de Conservación de la Ciudad de México]. We also want to thank the collaborative team SEDEMA-LANCIS for their support.

Author details

Yosune Miquelajauregui[1]* and Adela Madariaga-Fregoso[2]

1 Laboratorio Nacional de Ciencias de la Sostenibilidad, Instituto de Ecología, Universidad Nacional Autónoma de México (UNAM), México City, México

2 Facultad de Ciencias, Universidad Nacional Autónoma de México (UNAM), México City, México

*Address all correspondence to: yosune@iecologia.unam.mx

References

[1] Bojórquez-Tapia LA, Eakin H, Hernández-Aguilar B, Shelton R. Addressing complex, political and intransient sustainability challenges of transdisciplinarity: The case of the MEGADAPT project in Mexico City. Environmental Development. Jun 2021;**38**:100604

[2] Adger WN, Arnell NW, Tompkins EL. Successful adaptation to climate change across scales. Global Environmental Change. 2005;**15**:77-86

[3] Ford DJ, Ford-Berrang L, Paterson J. A systematic review of observed climate change adaptation in developed nations. Climatic Change. 2011;**106**:327-336

[4] Bartelet HA, Barnes ML, Cumming GS. Determinants, outcomes, and feedbacks associated with microeconomic adaptation to climate change. Regional Environmental Change. 2022;**22**:59

[5] Folke C, Carpenter SR, Walker B, Cheffer M, Chapin T, Rockstrom J. Resilience thinking: Integrating resilience, adaptability and transformability. Ecology and Society. 2010;**15**:20

[6] Berrang-Ford LM, Ford JD, Patterson J. Are we adapting to climate change? Global Environmental Change. 2021;**21**:25-33

[7] Wise RM, Fazey I, Stafford Smith M, Park SE, Eakin HC, Archer Van Garderen ERM, et al. Reconceptualizing adaptation to climate change as part of pathways and response. Global Environmental Change. 2014;**28**:325-336

[8] Scoones I, Stirling A, Abrol D, Atela J, Charli-Joseph L, Eakin H, et al. Transformations to sustainability: Combining structural, systemic and enabling approaches. Current Opinion in Environmental Sustainability. 2020;**42**:65-75

[9] Fazey I et al. Ten essentials for action-oriented and second order energy transitions, transformations and climate change research. Energy Research & Social Science. 2018;**40**:54-70

[10] Kates RW, Reavis WR, Willbanks T. Transformational adaptation when incremental adaptations to climate change are insufficient. Proceedings of the National Academy of Science. 2012;**109**:7156-7161

[11] Shi L, Moser S. Transformative climate change adaptation in the United States: Trends and prospects. Science. 2021;**372**(6549):eabc8054

[12] Moallemi EA, Kwakkel J, de Haan FJ, Bryan BA. Exploratory modeling for analyzing coupled human-natural systems. Global Environmental Change. 2020;**65**:102186

[13] Lempert RJ. Robust decision making (RDM). In: Marchau V, Walker W, Bloemen P, Popper S, editors. Decision Making Under Deep Uncertainty. Cham: Springer; 2019. pp. 23-50

[14] Bhave AG, Conway D, Dessai S, Stainforth DA. Barriers and opportunities for robust decision-making approaches to support climate change adaptation in the developing world. Climate Risk Management. 2016;**14**:1-10

[15] Eakin H, Parajuli J, Yogya Y, Hernández B, Manheim M. Entry points for addressing justice and politics in urban flood adaptation decision making. Current opinion in Environmental Sustainability. 2021;**51**:1-6

[16] Cash DW, Clark WC, Alcock F, Dockson NM, Eckley N, Guston DH. Knowledge systems for sustainable development. PNAS. 2003;**100**: 8086-8091

[17] Schneider F, Tribaldos T. Co-production of knowledge and sustainability transformations: A strategic compass for global research networks. Current Opinion in Environmental Sustainability. 2021;**49**:127-142

[18] Lemos MC et al. To co-produce or not to co-produce. Nature Sustainability. 2018;**1**(12):722-724

[19] Simpson NP, Mach KJ, Constable A, Hess J, Hogarth R, Howden M, et al. A framework for complex climate change risk assessment. One Earth [Internet]. 2021;**4**(4):489-501

[20] Reed PM et al. Multisector dynamics: Advancing the Science of Complex Adaptive human-earth-systems. 2022

[21] Fedele G, Donatti C, Harvey C, Hannah L, Hole DG. Transformative adaptation to climate change for sustainable socio-ecological systems. Environmental Science and Policy. 2019;**101**:116-125

[22] Bojórquez-Tapia LA, Eakin H, Reed P, Miquelajauregui Y, Grave I, Merino-Benitez T, et al. Unveiling uncertainties to enhance sustainability transformations in infrastructure decision-making. Current Opinion in Environmental Sustainability. 2022;**55**(101172):101172

[23] Klenk N, Fiume A, Meehan K, Gibbes C. Local knowledge in climate adaptation research: Moving knowledge frameworks form extraction to co-production. Wiley Interdisciplinary Reviews: Climate Change. 2017;**8**:1-15

[24] Hung H-H, Yang C-Y, Chien C-Y, Liu Y-C. Building resilience: Mainstreaming community participation into integrated assessment of resilience to climatic hazards in metropolitan land use management. Lan Use Policy. 2016;**50**:48-58

[25] Fatoric S, Seekamp E. Knowledge co-production in climate adaptation planning of archeological sites. Journal of Coastal Conservation. 2019;**23**:689-698

[26] Miquelajauregui Y, Bojórquez-Tapia LA, Eakin H, Gómez-Priego P, Pedroza D. Challenges and opportunities for universities in building adaptive capacities for sustainability: Lessons from Mexico, Central America and the Caribbean. Climate Policy. 2022;**22**(5):637-651

[27] Schipper ELF. Maladaptation: When adaptation to climate change goes very wrong. OneEarth-Primer. 2020;**3**:409-414

[28] Haasnoot M, Kwakkel JH, Walker W, ter Maat J. Dynamic adaptive policy pathways: A method for crafting robust decision for a deeply uncertain world. Global Environmental Change. 2013;**23**:485-498

[29] Pereira L, Kuiper JJ, Selomane O, Aguiar APD, Asrar GR, Bennett EM, et al. Advancing a toolkit of diverse future approaches for global environmental assessments. Ecosystems and People. 2021;**17**:191-204

[30] Norstrôm AV et al. Principles of knowledge co-production in sustainability research. Nature Sustainability. 2020;**3**:182-190

[31] Wyborn CA. Connecting knowledge with action through coproductive capacities: Adaptive governance and

connectivity conservation. Ecology and Society. 2015;**20**:11

[32] Kirchhoff CJ, Lemos MC, Dessai S. Actionable knowledge for environmental decision making: Broadening the usability of climate science. Annual Review of Environment and Resources. 2013;**38**:393-414

[33] Chambers JM, Wyborn K, Ryan M. Six models of co-production for sustainability. Nature Sustainability. 2021;**4**:983-996

[34] Lemos MC, Morehouse BJ. The co-production of science and policy in integrated climate assessments. Global Environmnetal Change. 2005;**15**:57-68

[35] Bremer S, Meisch S. Co-production in climate change research: Reviewing different perspectives. WIREs Clim Change. 2017

[36] Islam N, Winkel J. Climate change and social inequality. DESA Working Paper No. 152. 2017

[37] Doss-Gollin J, Farnham DJ, Steinschneider S, Lall U. Robust adaptation to multiscale climate variability. Earth's Future. 2019;7:734-747

Chapter 13

Biostimulants Application: An Innovative Approach to Food Security under Drought Stress

Muhammad Amjad Bashir, Qurat-Ul-Ain Raza, Abdur Rehim, Muhammad Umair Sial, Hafiz Muhammad Ali Raza, Saba Ali, Muhammad Ijaz, Faiz-Ul-Hassan and Yucong Geng

Abstract

Climate change is a global threat to food security as it causes various biotic and abiotic stresses that adversely disturb agriculture production. With an increase in the worldwide population, the demand for food has also arisen. It is an immediate challenge for the scientific community to introduce an innovative tool to achieve food security with quality plant production and develop tolerance against abiotic stresses, specifically drought. Genetically modifications are effective and time-consuming, while biostimulants are in/organic substances with the potential to support plant development under stress conditions. This chapter focuses on the impacts of climate change on agriculture, challenges for agriculture sustainability and food security, the interrelationship between drought, climate change and food security, the potential role of biostimulants against drought, future aspects and challenges due to climate change specifically drought, and food security challenges. Various studies reported that the application of biostimulants results in enhancement of crop productivity and mitigates the harmful effects due to climate change. To ensure the quality of chapter, we collected references from well-reputed international journals using keywords ""biostimulants," "drought," "food security," "agriculture sustainability," and "climate change." In conclusion, biostimulants has a potential to address adverse environmental conditions without affecting crop quality and yield loss.

Keywords: abiotic stress, biostimulants, climate change, food security, drought stress

1. Introduction

The world population has doubled since 1960, resulting in increased food demand and agricultural production [1]. World food production needs to be increased to 70% to feed this population in 2050 [2]. Due to high warming temperatures and precipitation shifts, agriculture has faced a massive yield decline, especially in lower latitudes [3]. Global warming speeds up crop development and reduces the maturity time and photo-assimilation period thus disturbs crop yield and quality. Moreover,

IntechOpen

such climatic conditions are more optimistic for pests. During the growing season, the climatic water balance has been projected to increase negatively, leading to water shortage [4]. Furthermore, it also affects the stomatal closure, cell damage, delayed seed germination, disturbed structure and functionality of cell membranes, enzyme inactivity, and interferences on protein synthesis, ultimately damaging crop productivity [5].

Drought is another environmental hazard described as a long period of reduced precipitation occurring in almost all climatic zones, including low and high rainfall [6]. Crop productivity and yield depend upon irrigation management, water quality, and regimes. At the same time, water shortage disturbs gaseous exchange, photosynthetic activity, evapotranspiration, stomatal closure, and nutrient uptake and consequently affects plant biomass [1]. Drought stress events are mainly associated with low rainfall and high soil evaporation due to high-temperature events, dry wind, and high light intensity. Scientists and agriculturists have introduced various drought-resistance induction strategies to cope with drought and other global warming impacts [7].

Biostimulants are substances having the potential to improve nutrient use efficiency and uptake, develop resistance against biotic and abiotic stresses, and improve quality characteristics when applied to plants [8]. Moreover, biostimulants vary in composition depending on the material used in their preparation. It enhances plant growth and nutrition when applied in minute quantities; therefore, it should not be termed as fertilizers and other soil amendments used in considerable amounts to achieve the required yield [9]. Studies reported that paramylon [10], commercial *Ascophyllum nodosum* extracts [11], and exogenous application of melatonin developed drought resistance and improved tomato quality [12]. Under drought conditions, mint quality and quantity were improved using biostimulants [13].

The objectives of this chapter were to identify the impacts of climate change on agriculture sustainability and the role of biostimulants in drought. Climate change has adversely impacted the agriculture sustainability while the biostimulants are gaining popularity due to its potential in addressing abiotic stresses. Therefore, the chapter focuses on provided the alarming signals about climate change and how farmers and other landholders can use biostimulants to achieve food security. The chapter consists of eight sections, which include the introduction (Section 1), methodology (Section 2), impacts of climate change on agriculture, both biotic and abiotic stresses (Section 3), challenges for agriculture sustainability and food security (Section 4), the interrelationship between drought, climate change, and food security (Section 5), potential role of biostimulants against drought (Section 6), future aspects and challenges (Section 7), and conclusion (Section 8), respectively.

2. Methodology

To ensure the quality of the chapter, we reviewed researched articles, review articles, books, and scientific reports only indexed by Scopus, Web of Science, Science Direct, and Google Scholar. We targeted specific keywords including "biostimulants," "drought," "food security," "agriculture sustainability," and "climate change." The articles published in well-reputed journals were studied. Moreover, the articles not related to objectives of the chapter were eliminated. The data and information collected were transformed into table and figures.

3. Impact of climate change on agriculture

Overwhelming environmental changes have harmed agricultural production, human health, and natural systems [14]. Agriculture and climate change are linked in numerous ways since climate change is the leading driver of biotic and abiotic pressures that have detrimental effects on agriculture in an area. Concerns over the stability of the worldwide environment have led to an increase in food demand in tandem with the rapid growth of the global population. Agriculture productivity is greatly affected by water availability, air pollution, and soil quality [15]. Climate change impacts land and agriculture in many ways, including changes in yearly rainfall, heat waves, average temperature, weeds, insects, microbes, and atmospheric CO_2 or ozone level.

3.1 Effects on abiotic factors

3.1.1 Temperature

Temperature affects the growth and development of plants depending on the crop being grown [16]. Climate change reduces rainfall, wind speed, and snow cover due to rising temperatures and shortens the growing season for plants, affecting crop quality and agricultural productivity [7]. The causes of temperature rise can be traced back to global warming, which varies from place to region. In the future, developing countries will be more vulnerable, which may lead to a rise in food insecurity in the region. According to a study on the effects of frost and extreme temperatures on wheat production (*Triticum aestivum* L), frost produced unfruitfulness and abortion of created grains, while excessive heat resulted in a reduction in the number of grains formed during the filling period of the grain [17]. The high-temperature effects on a pearl millet were studied by [18], and the researchers identified sensitive stages of the plant's growth process. This research assessed temperature thresholds, genetic diversity, and pollen fertility.

Moreover, the high temperature reduces pollen germination and seed production. This also impacts pollen and pistil fertility [18, 19]. Due to the effects of climate change on agricultural production, climatic variance threatens crop production patterns, causing food insecurity.

3.1.2 Drought and rainfall

One essential abiotic variable that reduces the number of agricultural products harvested worldwide is drought [20]. It influences not only the growth of the crops but also the yield value. In an experiment on miscanthus for biofuel generation, drought treatment lowered plant weight by 45% and affected biomass composition and cell wall structural stiffness [21]. Due to the distribution and pattern of precipitation in tropical regions, the water content of the soil in these regions varies significantly [22]. This means plant water in the soil is dwindling. In addition, research to investigate the impact of precipitation timing on rainforest and grassland in the United States found that plant-usable soil water content depends on precipitation [23]. In other words, when rainfall distribution is uneven, the soil water content decreases, producing stress on plants in afflicted locations. This is frequently the outcome of climate change.

3.1.3 Waterlogging/flooding

Climate change has disrupted the hydrological cycle, reducing or impairing agricultural growth in many parts of the world. As a result, waterlogging significantly impacts agricultural productivity, particularly on flatland or areas near rivers [24]. Heavy rainfall in the area is the primary source of waterlogging, although irrigation canal leaks and clean surface drainage can also contribute. Soil compaction increases, and the amount of accessible O_2 for plant cells decreases because the diffusion process of O_2 is sluggish in ponding water [25]. Consequently, anaerobic bacteria release iron ions, manganese ions, and sulfide in large quantities because oxygen is scarce. Physiological and morphological changes occur in crops that are waterlogged [25]. In reaction to waterlogging, a plant's stomata closes, which affects gas exchange and water uptake, as well as anaerobic conditions in the rhizosphere, harms the plant's ability to absorb water [24].

3.1.4 Salinity

According to [19], worldwide salinity affects crop yield and food supplies. Since salt-sensitive crops (wheat and rice) are grown worldwide (F.A.O., 2015), salinity must be addressed soon. According to [26], rice is one of the most widely cultivated crops since it is a crucial source of sustenance for nearly all of humanity. Salinity is a stressor in dry and semiarid environments when evapotranspiration exceeds rainfall, resulting in insufficient rain to filter soluble salts from the root zone. The salinity stressor inhibits plants' capacity to absorb nutrients and water from the soil, stunting their development; salt deposits in the transpiration stream harm leaf cells, causing leaf burn; it also alters enzyme activity within the plant [27].

3.2 Effects on biotic factors

3.2.1 Livestock

The focus of the cattle industry during the past quarter-century has been on enhancing production, altering the environment, and enhancing nutritional management rather than improving stress resistance. This method substantially boosted the output of domestic animals but also increased their susceptibility to hot surroundings. The modes by which domestic animals adjust to environmental changes are crucial to their survival, but they frequently have a detrimental impact on the productivity and profitability of livestock systems [28]. Heat stress has damaging effects on the health and welfare of animals. The direct and indirect impacts of heat stress on the health of farm animals in hot environments. Increased temperatures, frequency, and severity of heat waves are the primary causes of the direct consequences. These climatic circumstances can harm the health of cattle by generating metabolic changes, oxidative stress, and immunological suppression, which can lead to illnesses and mortality. Indirect consequences include changes in the availability and quality of feedstuffs and drinking water and the survival and redistribution of diseases and/or their vectors [29].

3.2.2 Aquaculture and fisheries

Aquatic ecosystems are essential to the global environment. In addition to being crucial contributors to biodiversity and ecological production, wetlands offer several benefits to human populations, such as water for drinking and irrigation, recreational activities, and habitat for commercially significant fisheries [30]. Marine fisheries contribute significantly to people's and society's well-being, particularly in the tropics, where coastal populations rely on fisheries for food security, livelihoods, economic growth, and culture [31]. Fisheries are becoming increasingly vulnerable to changes in the physical and biogeochemical properties of the ocean (such as warming, sea-level rise, deoxygenation, acidification, and altered nutrient concentrations) caused by rising concentrations of anthropogenic greenhouse gases, particularly CO_2 [32]. The distribution, abundance, and reproduction of fish and invertebrate species are also being affected by physical and biogeochemical stresses via ecosystems, directly and indirectly impacting fisheries productivity [33]. By the year 2050, climate change may cause 10–40% of species that are appropriate for marine aquaculture to become extinct in the tropics and subtropics [34].

3.2.3 Insect pests

Pests are a key biotic component also affected by climate change and weather disturbances. Temperature increases have an immediate impact on pest reproduction, survival, dissemination, population dynamics, and interactions between pests, the environment, and natural enemies. As a result, it is critical to monitor pest presence and abundance since the conditions of their occurrence might change quickly [35]. The effect of climate change on arthropod extinction rates is between 100 and 1000 times more prominent than in the past, with 45–275 species becoming extinct daily. A temperature increase of 6°C would result in the extinction of several species, including humans. In North America and Europe, bumblebee populations have decreased by 46% and 17%, respectively, because of extreme temperatures caused by climate change, compared with the base period of 1901–1974 [36]. Climate change produces new ecological niches that allow insect pests to develop and proliferate in new geographic locations and migrate from one region to another. Due to the changing environment, farmers should expect to encounter new and significant insect issues in the following years. The spread of agricultural pests across physical and political borders threatens food security and is a global issue shared by all nations and regions.

The physiology of insects is extremely sensitive to variations in temperature; as a rule, their metabolic rate will almost double for every 10°C increase in temperature [36]. Populations of whiteflies are primarily influenced by environmental conditions such as temperature, precipitation, and humidity. Whitefly population growth is favorably associated with high temperature and humidity [37]. Increased atmospheric CO_2 levels can impact the distribution, number, and productivity of insects that feed on plants. Such increases may impact insect pests' growth, fertility, consumption rates, and population densities [38]. Climate change is expected to affect the amount, distribution, and seasonal timing of pests and their natural enemies, changing biological control activities [39]. Aphids are handled by natural enemies such as parasitic wasps and ladybirds. These species may react differently to temperature changes due to global warming [40].

4. Challenges for agricultural sustainability and food security

Agriculture, covering more than 40 of the world's land, is the sole food provider for human beings and animals [41]. It also plays a vital role in most countries' economic growth [42]. Furthermore, on an average basis, about 77% of the per capita energy requirements in the world are also fulfilled through crop base food products, while 23% of the remaining food comes from other sources, including meat, egg, and milk [43]. Consequently, the nutritional demand of the increasing population is one of the basic needs that can only be attained by increasing agricultural production [44]. So, in developing economies, there is a direct relationship between employment generation and poverty eradication with the progress in the agricultural sector [45]. However, this sector faces many problems but is also developing constantly through adopting various measures to handle these numerous challenges.

Climate change is likely to have harsh effects on various influences, including water resources, coastal regions, agriculture, human liveliness, food security, ecosystems and biodiversity. According to [46], climate change may predict about a 30% decrease in the yield of different crops in some parts of south and central Asia. Modifications in agricultural practices increased industrial products because of the global inevitability to assure access to nutrition on behalf of the increasing population along with the assimilation of markets and globalization (**Figure 1**) [47].

The world's food demand is increasing daily because of the increasing population. So, adequate food security is a critical problem [48]. According to the World Food Summit, food security can be defined as "when all the people in the world, any time, have economically and physically access to adequate, enough, and safe nutritious

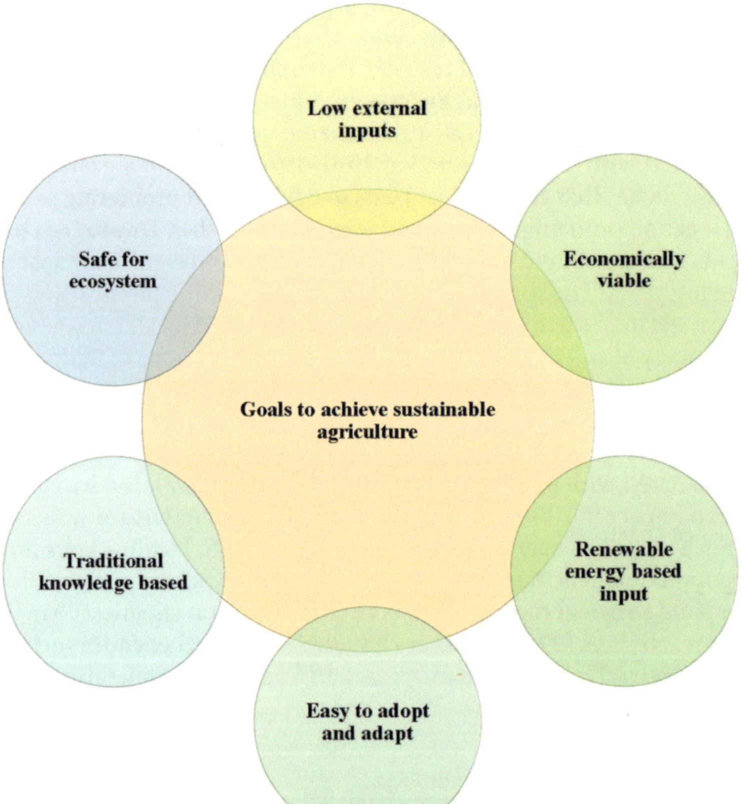

Figure 1.
Need and goals to achieve agricultural sustainability.

to fulfil their dietary requirements along with food partialities to spent healthy and active life" [49]. Food security has four main components: stability, availability, utilization, and access to food [50]. Nowadays, many challenges are faced by food security, starting from the application of various fertilizers, such as phosphorous, potassium, and nitrogen, to agricultural lands [51], then the deteriorating water tables, increasing temperatures, and an abrupt increase in population as well as consumption progression [52]. So, the two main challenges to food security are global inequity and entrepreneurial forms of production and distribution.

To cope with all these challenges, there is a need to make progress in scientific knowledge along with novel agricultural technologies that allow the communal urban agriculture practices to change with the present advanced urban agriculture. There are two types of urban agriculture, controlled environment agriculture and uncontrolled environment agriculture. According to [53], controlled environment agriculture contains agriculture practices related to environmental optimization, usually in combination with immediate urban assemblies, such as uses of greenhouse, internal farming, vertical farming, and building assimilated agriculture. While, in uncontrolled environment agriculture, vegetable farming is done in open space instead of a greenhouse and contains different types of gardens, including rooftop and community gardens that are usually indicated to play a vital role in food security globally.

5. The interrelationship between drought, climate change, and food security

Drought causes many physiological and molecular disorders in plants through excessive production of reactive oxygen species (ROS). It negatively influences the morphophysiological traits, including plant height, leaf area, relative water content, stomatal oscillation, chlorophyll contents, osmotic potential, and leaf water potential in crops [54]. Drought damages the photosynthetic process and causes stomatal closure. The reduced photosynthesis due to stomatal closure is reported to limit the supply of CO_2 [55]. Food security is dependent on social, economic, and climatic factors. Climatic extremes, particularly droughts and floods, affect the state of food security in Africa [56].

One of the key predictions of climate change is that in some regions, droughts are likely to increase in frequency and severity. This will have significant implications for the long-term viability of plant populations, especially where water availability plays a key role in delineating species ranges [57]. The human influence on the earth's climate is becoming increasingly prominent. Climate observations prove the existence of a global warming trend: global average temperature has increased by 0.88°C since 1900 [58], and the 12 hottest years observed globally since 1880 all occurred between 1990 and 2005. The climate changes will also have associated consequences for biotic (frequency and effects of pest and disease outbreaks) and abiotic disturbances (changes in fire occurrence, changes in wind storm frequency and intensity) with substantial implications for forest ecosystems [59].

Public awareness of the importance of extreme climatic events is growing [60]. While longer-term climatic reconstructions suggest that the occurrence and impacts of such climatic events are not new (e.g., Acuna-Soto et al., 2005; Benson et al., 2007), there is now a growing concern that anthropogenic global warming could increase the severity and frequency of extreme climatic events in the future [60]. From an ecological perspective, climate change also represents a major threat to global biodiversity

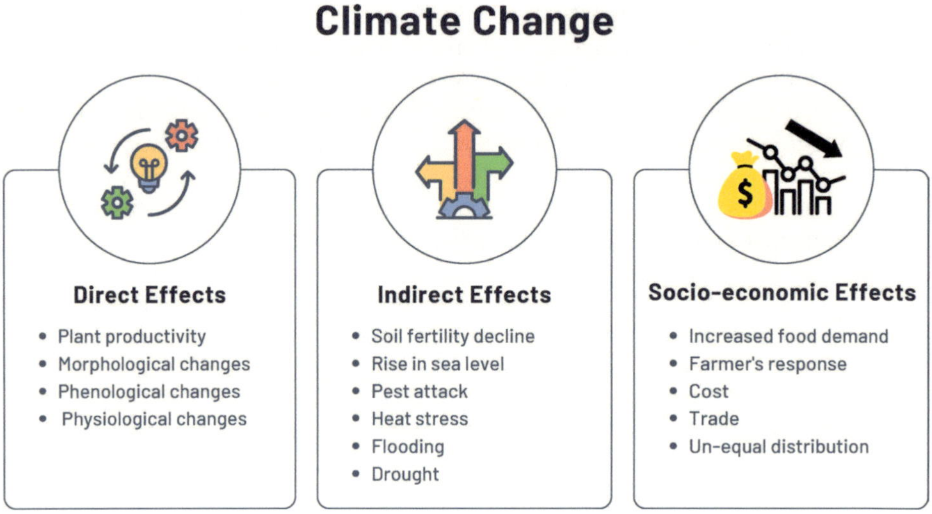

Figure 2.
Direct, indirect, and socioeconomic effects of climate change.

conservation. Indeed, it is argued that in the absence of rapid implementation of strategies to reduce global greenhouse gas emissions and other processes, many of the earth's biota will likely be committed to extinction [61]. By disturbing plant populations and derived ecosystem functions that depend on complex species interactions, extreme climatic events are likely to play a central mechanistic role, which influences food security [62]. The relationship between drought and agriculture is particularly important as 84% of the damage and losses caused by droughts relate to the agriculture sector [63]. Globally, droughts and extreme heat significantly reduced national cereal production by 9–10%, whereas the analysis could not identify flood and harsh cold effects [64]. Drought disasters in Indonesia mainly occur in Java and Madura since both islands have higher risk and vulnerability than other islands (**Figure 2**) [65].

Whereas the outcomes of abiotic stresses on crop yield are hard to calculate accurately, it is believed that abiotic stresses substantially influence crop production depending upon the extent of damage to the total area under cultivation. In future, the productivity of the major crops is estimated to drop in many countries due to global warming, water shortage, and other environmental impacts [66]. To achieve food security, water must be available at the right place, at the right time, in the right quantity, and be of the right quality. Water-related disasters negatively affect agriculture and crop production, threatening food security [67].

6. Potential role of biostimulants against drought

Drought stress impacts the agriculture sector and can adversely affect the plant's morphological, physiological, biochemical, and molecular changes, leading to reduced plant growth and productivity [68]. Water stress also results in increased reactive oxygen species (ROS) accumulation in the plant, which disturbs different cellular mechanisms, including enzyme inhabitation, protein reductase, DNA, RNA, and membrane lipid peroxidation damage leading to cell death [69]. It also disturbed the stomatal closure, enzyme activity, photosynthesis process, and transpiration rates [70].

Biostimulants can boost crop toleration against drought stress and improve crop productivity [9]. Moreover, it can potentially increase the growth and yield of plants in farming by developing tolerance against drought stresses, enhancing water retention capability, root strengthening and development, and improving nutrient and water use efficiency [71]. The impact of biostimulants highly depends on time and rate of application, abiotic environmental factors, and crop variety [72]. Studies reported that using pollen grain extract as biostimulant improved the growth and essential oil productivity of *Ocimum basilicum* under drought stress conditions [73].

In addition, arbuscular mycorrhizal fungi, plant-growth-promoting rhizobacteria, and local green compost improve date palm productivity under a water stress environment [74]. *Ascophyllum nodosum* extract biostimulants also developed drought tolerance in tomato plants [11]. Under mild drought stress, the application of *A. nodosum* seaweed extract (SWE) enhanced spinach growth [75]. While foliar-applied (SWE) improved bean yield cultivars under irrigation regimes by changing fatty acid and biochemical profiles [76]. Further benefits of biostimulants and plant responses under different conditions are given in **Table 1**.

7. Future aspects and challenges

To make water available for human consumption and irrigation, an adequate water management system must be adopted worldwide for sustainable agriculture and profitable activity of water, especially in arid and semiarid regions [83]. Biostimulants have been reported to address drought stress while maintaining crop yield and productivity [9]. However, the response of different crop cultivars and the long-term impacts of biostimulants need to be studied. Moreover, multiple sources of biostimulants should be adopted after identifying their synergistic effects rather than relying on a single basis. Literature reported the research conducted at tunnel farms, experimental greenhouse areas, and hydroponic conditions. However, field experiments should be performed to find the potential of biostimulants in real-life situations. In addition, the use of biostimulants can also provide environmental benefits. It can be used to tackle food demand by improving productivity under environment stress as well as utilization of waste products to prepare biostimulants for environment sustainability.

Furthermore, the effectiveness of biostimulants in normal environmental conditions and their comparison with chemical fertilizers can also give new insights. The use of biostimulants for sustainable agriculture to reduce agrochemical products (such as chemical fertilizers and pesticides) can be studied. Moreover, the high demand for biofertilizers and biopesticides can be addressed using natural biostimulants, including seaweed, pollen grain, moringa leaf, and many other extracts. Therefore, new and economical biostimulants with different compositions should be introduced into the market along with their specifications to specific crop/cultivar, application method, time, and dose, and their effectiveness toward specific stress conditions should be mentioned. Biostimulants vary in nature depending upon their composition and the material from which they are derived. Due to the uniqueness of every type of biostimulant, the mechanism behind its activity and performance is still not identified; thus, further studies are required in this regard because it can help scientists identify more benefits of biostimulants.

Biostimulant	Crop	Stress	Results	Reference
Saprophytic fungi (*Trichoderma harzianum* ALL-42)	Common bean (*Phaseolus vulgaris*)	Pathogenic stress	Improved shoot biomass production and number of lateral shoots, module plant's metabolism and triggers its defense response	[77]
Plant and seaweed extract	Baby spinach (*Spinacia oleracea*)	None	Improved plant growth, quality, and yield	[78]
Fresh seaweed extract of *Ascophyllum nodosum*	Lettuce	Potassium deficiency	Increased the quality of cut lettuce	[79]
Alfalfa (*Medicago sativa*) and red grape (*Vitis vinifera*) derived biostimulants	*Capsicum chinensis*	None	Promoted plant growth and the production of secondary metabolites	[80]
Salicylic acid, beeswax waste and liquorice extract	Sesame	Drought stress	Mitigate drought stress and oxidative damages, regulate osmoprotectants and antioxidant defense system and improve sesame productivity.	[81]
A. nodosum seaweed extracts	Spinach	Drought stress	Enhanced gas exchange through reduction of stomatal closure, resulting in increased plant resistance to water stress.	[75]
ERANTHIS®® (seaweed A. nodosum, Laminaria digitata, and yeast-based extracts)	Tomato	Drought stress	Mitigate water stress	[82]
Biostimulants, Vitamin B12, and CoQ10	Red radish	None	Improved fresh and dry weights of roots and shoots	[72]
Glycine, lysine, aspartic acid, vitamin B complex, and moringa leaf extract	Radish	None	Improved morpho-physiology properties and yield	[8]
A.M.F. consortium (A.M.F.), indigenous PGPR (B), and local green compost	Date palm	Drought stress	Mitigated drought stress, improved plant biomass and phosphorus uptake, and boosted plant-water relationship	[74]

Table 1.
Benefits of biostimulants and plant responses.

8. Conclusion

High food demand, climate change, increasing incidences of weather extremities, and other biotic and abiotic stresses have created severe pressure on crops. Conventional agriculture practices are challenging to maintain in such scenarios.

Therefore, farmers and scientists are facing the challenge of finding an appropriate approach to tackle environmental stresses and maintain crop productivity. Biostimulants are an innovative tool to address the issue without adversely affecting crop quality and yield loss. We studied the latest research in this aspect, and biostimulants have reported positive responses against adverse environmental conditions. However, further studies are required to identify the crop response in different areas of the globe and its long-term potential under other stress conditions.

Author details

Muhammad Amjad Bashir[1†], Qurat-Ul-Ain Raza[2†], Abdur Rehim[1,2*], Muhammad Umair Sial[3], Hafiz Muhammad Ali Raza[1,2], Saba Ali[1], Muhammad Ijaz[1], Faiz-Ul-Hassan[1] and Yucong Geng[4*]

1 College of Agriculture, Bahauddin Zakariya University, Bahadur Sub-Campus, Layyah, Pakistan

2 Department of Soil Science, FAS&T, Bahauddin Zakariya University, Multan, Pakistan

3 Department of Entomology, University of Agriculture Faisalabad, Pakistan

4 KOYO Star Agriculture Technology Co., LTD, Beijing, China

*Address all correspondence to: abdur.rehim@bzu.edu.pk and 173387338@qq.com

† These authors contributed equally.

References

[1] Del Buono D. Can biostimulants be used to mitigate the effect of anthropogenic climate change on agriculture? It is time to respond. Science of the Total Environment. 2021;**751**:141763

[2] van Dijk M, Morley T, Rau ML, Saghai Y. A meta-analysis of projected global food demand and population at risk of hunger for the period 2010-2050. Nature Food. 2021;**2**:494-501. DOI: 10.1038/s43016-021-00322-9

[3] Froehlich HE, Koehn JZ, Holsman KK, Halpern BS. Emerging trends in science and news of climate change threats to and adaptation of aquaculture. Aquaculture. 2022;**549**:737812

[4] Bisbis MB, Gruda N, Blanke M. Potential impacts of climate change on vegetable production and product quality – A review. Journal of Cleaner Production. 2018;**170**:1602-1620

[5] Fahad S, Bajwa AA, Nazir U, Anjum SA, Farooq A, Zohaib A, et al. Crop production under drought and heat stress: Plant responses and management options. Frontiers in Plant Science. 2017;8:1147

[6] Orimoloye IR, Belle JA, Orimoloye YM, Olusola AO, Ololade OO. Drought: A common environmental disaster. Atmosphere (Basel). 2022;**13**:111

[7] Seleiman MF, Al-Suhaibani N, Ali N, Akmal M, Alotaibi M, Refay Y, et al. Drought stress impacts on plants and different approaches to alleviate its adverse effects. Plants. 2021;**10**:1-25

[8] Raza Q-U-A, Bashir MA, Rehim A, Ejaz R, Raza HMA, Shahzad U, et al. Biostimulants induce positive changes in the radish morpho-physiology and yield. Frontiers in Plant Science. 2022;**0**:2475. DOI: 10.3389/FPLS.2022.950393

[9] Bashir MA, Rehim A, Raza Q-U-A, Muhammad Ali Raza H, Zhai L, Liu H, et al. Biostimulants as plant growth stimulators in modernized agriculture and environmental sustainability. In: Technology in Agriculture [Working Title]. London, UK: IntechOpen; 2021 ISBN 978-1-83881-922-4

[10] Barsanti L, Coltelli P, Gualtieri P. Paramylon treatment improves quality profile and. Agronomy. 2019;**9**:1-16

[11] Goñi O, Quille P, O'Connell S. Ascophyllum nodosum extract biostimulants and their role in enhancing tolerance to drought stress in tomato plants. Plant Physiology and Biochemistry. 2018;**126**:63-73. DOI: 10.1016/j.plaphy.2018.02.024

[12] Karaca P, Cekic FÖ. Exogenous melatonin-stimulated defense responses in tomato plants treated with polyethylene glycol. International Journal of Vegetable Science. 2019;**25**:601-609. DOI: 10.1080/19315260.2019.1575317

[13] Elansary HO, Mahmoud EA, El-Ansary DO, Mattar MA. Effects of water stress and modern biostimulants on growth and quality characteristics of mint. Agronomy. 2020;**10**:6. DOI: 10.3390/agronomy10010006

[14] Araújo MB, Rahbek C. How does climate change affect biodiversity? Science (80-.). 2006;**313**:1396-1397

[15] Noya I, González-García S, Bacenetti J, Fiala M, Moreira MT. Environmental impacts of the cultivation-phase associated with

agricultural crops for feed production. Journal of Cleaner Production. 2018;**172**:3721-3733. DOI: 10.1016/j.jclepro.2017.07.132

[16] Hatfield JL, Prueger JH. Temperature extremes: Effect on plant growth and development. Weather and Climate Extremes. 2015;**10**:4-10. DOI: 10.1016/j.wace.2015.08.001

[17] Barlow KM, Christy BP, O'Leary GJ, Riffkin PA, Nuttall JG. Simulating the impact of extreme heat and frost events on wheat crop production: A review. Field Crops Research. 2015;**171**:109-119. DOI: 10.1016/J.FCR.2014.11.010

[18] Djanaguiraman M, Perumal R, Ciampitti IA, Gupta SK, Prasad PVV. Quantifying pearl millet response to high temperature stress: Thresholds, sensitive stages, genetic variability and relative sensitivity of pollen and pistil. Plant, Cell & Environment. 2018;**41**:993-1007. DOI: 10.1111/pce.12931

[19] Dhankher OP, Foyer CH. Climate resilient crops for improving global food security and safety. Plant, Cell & Environment. 2018;**41**:877-884

[20] Tuberosa R, Grillo S, Ellis RP. Unravelling the genetic basis of drought tolerance in crops. In: Abiotic Stresses in Plants. Netherlands: Springer; 2003. pp. 71-122

[21] van der Weijde T, Huxley LM, Hawkins S, Sembiring EH, Farrar K, Dolstra O, et al. Impact of drought stress on growth and quality of miscanthus for biofuel production. GCB Bioenergy. 2017;**9**:770-782. DOI: 10.1111/gcbb.12382

[22] Smith MD. An ecological perspective on extreme climatic events: A synthetic definition and framework to guide future research. Journal of Ecology. 2011;**99**:656-663

[23] Zeppel MJB, Wilks JV, Lewis JD. Impacts of extreme precipitation and seasonal changes in precipitation on plants. Biogeosciences. 2014;**11**:3083-3093. DOI: 10.5194/bg-11-3083-2014

[24] Aldana F, García PN, Fischer G. Effect of waterlogging stress on the growth, development and symptomatology of cape gooseberry (Physalis peruviana L.) plants. Revista de la Academia Colombiana de Ciencias Exactas, Físicas y Naturales. 2014;**38**:393. DOI: 10.18257/raccefyn.114

[25] Ashraf M, Mehmood S. Effects of waterlogging on growth and some physiological parameters of four brassica species. Plant and Soil. 1990;**121**:203-209. DOI: 10.1007/BF00012313

[26] Food and Agricultural Organisation of the United Nations. The Future of Food and Agricultural Trends and Challenges. Rome: Food and Agricultural Organisation; 2015

[27] Fowler DN, King SL, Weindorf DC. Evaluating abiotic influences on soil salinity of inland managed wetlands and agricultural croplands in a semi-arid environment. Wetlands. 2014;**34**:1229-1239. DOI: 10.1007/s13157-014-0585-3

[28] Bernabucci U. Climate change: Impact on livestock and how can we adapt. Animal Frontiers. 2019;**9**:3-5

[29] Lacetera N. Impact of climate change on animal health and welfare. Animal Frontiers. 2019;**9**:26-31. DOI: 10.1093/af/vfy030

[30] Prakash S. Impact of climate change on aquatic ecosystem and its biodiversity: An overview. International Journal of Biological Innovations. 2021;**03**:312-317. DOI: 10.46505/ijbi.2021.3210

[31] Teh LCL, Pauly D. Who brings in the fish? The relative contribution of small-scale and industrial fisheries to food security in Southeast Asia. Frontiers in Marine Science. 2018;**5**:44. DOI: 10.3389/fmars.2018.00044

[32] Barange M, Bahri T, Beveridge MCM, Cochrane KL, Funge-Smith S, Poulain F. Impacts of climate change on fisheries and aquaculture. United Nations' Food and Agriculture Organization. 2015;12(4):628-635

[33] Laffoley D, Baxter JM, editors. Explaining ocean warming: Causes, scale, effects and consequences. Gland, Switzerland: IUCN; 2016

[34] Oyinlola MA, Reygondeau G, Wabnitz CCC, Cheung WWL. Projecting global mariculture diversity under climate change. Global Change Biology. 2020;**26**:2134-2148. DOI: 10.1111/gcb.14974

[35] Prakash A, Rao J, Mukherjee AK, Berliner J, Pokhare SS, Adak T, et al. Climate change: Impact on crop pests. Research Today. 2014;**2**:327-329

[36] Dukes JS, Pontius J, Orwig D, Garnas JR, Rodgers VL, Brazee N, et al. Responses of insect pests, pathogens, and invasive plant species to climate change in the forests of northeastern North America: What can we predict? Canadian Journal of Forest Research. 2009;**39**:231-248

[37] Pathania M, Verma A, Singh M, Arora PK, Kaur N. Influence of abiotic factors on the infestation dynamics of whitefly, Bemisia tabaci (Gennadius 1889) in cotton and its management strategies in North-Western India. International Journal of Tropical Insect Science. 2020;**40**:969-981. DOI: 10.1007/s42690-020-00155-2

[38] Fuhrer J. Agroecosystem responses to combinations of elevated CO2, ozone, and global climate change. Agriculture, Ecosystems and Environment. 2003;**97**:1-20

[39] Thomson LJ, Macfadyen S, Hoffmann AA. Predicting the effects of climate change on natural enemies of agricultural pests. Biological Control. 2010;**52**:296-306

[40] Hullé M, Coeur d'Acier A, Bankhead-Dronnet S, Harrington R. Aphids in the face of global changes. Comptes Rendus Biologies. 2010;**333**:497-503. DOI: 10.1016/j.crvi.2010.03.005

[41] Sah D, Devakumar AS. The carbon footprint of agricultural crop cultivation in India. Carbon Management. 2018;**9**:213-225. DOI: 10.1080/17583004.2018.1457908

[42] Bose B, Mondal S. Climate change and sustainable agriculture in context to seed priming and role of nitrate. Vegetos. 2013;**26**:192-204. DOI: 10.5958/j.2229-4473.26.2s.140

[43] Brevik EC, Burgess LC, editors. Soils and human health. Florida, USA: CRC Press; 2012. ISBN 9781439844557

[44] Singh R, Singh H, Raghubanshi AS. Challenges and opportunities for agricultural sustainability in changing climate scenarios: A perspective on Indian agriculture. Tropical Ecology. 2019;**60**:167-185

[45] Srivastava P, Singh R, Tripathi S, Raghubanshi AS. An urgent need for sustainable thinking in agriculture - an Indian scenario. Ecological Indicators. 2016;**67**:611-622. DOI: 10.1016/j.ecolind.2016.03.015

[46] Arora S, Bhatt R. Resource conservation technologies (R.C.T.s) for climate-resilient agriculture in the foothill of Northwest Himalayas.

In: Conservation Agriculture: An Approach to Combat Climate Change in Indian Himalaya. Singapore: Springer; 2016. pp. 71-111 ISBN 9789811025587

[47] Quintero-Angel M, González-Acevedo A. Tendencies and challenges for the assessment of agricultural sustainability. Agriculture, Ecosystems and Environment. 2018;**254**:273-281

[48] Santeramo FG, Carlucci D, De Devitiis B, Seccia A, Stasi A, Viscecchia R, et al. Emerging trends in European food, diets and food industry. Food Research International. 2018;**104**:39-47. DOI: 10.1016/j.foodres.2017.10.039

[49] Mah CL, Hamill C, Rondeau K, McIntyre L. A frame-critical policy analysis of Canada's response to the world food summit 1998-2008. Archives of Public Health. 2014;**72**:1-7. DOI: 10.1186/2049-3258-72-41

[50] Briones Alonso E, Cockx L, Swinnen J. Culture and food security. Global Food Security. 2018;**17**:113-127

[51] Cordell D, Drangert JO, White S. The story of phosphorus: Global food security and food for thought. Global Environmental Change. 2009;**19**:292-305. DOI: 10.1016/j.gloenvcha.2008.10.009

[52] Godfray HCJ, Beddington JR, Crute IR, Haddad L, Lawrence D, Muir JF, et al. Food security: The challenge of feeding 9 billion people. Science (80-.). 2010;**327**:812-818

[53] Al-Kodmany K. The vertical farm: A review of developments and implications for the vertical city. Buildings. 2018;**8**:24

[54] Sammar Raza MA, Saleem M, Khan I, Jamil M, Ijaz M, Khan MA. Evaluating the drought stress tolerance efficiency of wheat (Triticum aestivum l.) cultivars. Russian Journal of Agricultural and Socio-Economic Sciences. 2012;**12**:41-46. DOI: 10.18551/rjoas.2012-12.04

[55] Li M, Yang Y, Raza A, Yin S, Wang H, Zhang Y, et al. Heterologous expression of Arabidopsis thaliana rty gene in strawberry (Fragaria × ananassa Duch.) improves drought tolerance. BMC Plant Biology. 2021;**21**:1-20. DOI: 10.1186/s12870-021-02839-4

[56] Scoones I. Hazards and opportunities: Farming livelihoods in dryland Africa. Lessons from Zimbabwe. London, UK: Zed Books; 1996. ISBN 1856493539

[57] Alizadeh V, Shokri V, Soltani A, Yousefi MA. Effects of climate change and drought-stress on plant Physiologya. International Journal of Advanced Biological and Biomedical Research. 2014;**2**:468-472

[58] Hansen J, Sato M, Ruedy R, Lo K, Lea DW, Medina-Elizade M. Global temperature change. Proceedings of the National Academy of Sciences of the United States of America. 2006;**103**:14288-14293. DOI: 10.1073/pnas.0606291103

[59] Lindner M, Maroschek M, Netherer S, Kremer A, Barbati A, Garcia-Gonzalo J, et al. Climate change impacts, adaptive capacity, and vulnerability of European forest ecosystems. Forest Ecology and Management. 2010;**259**:698-709. DOI: 10.1016/j.foreco.2009.09.023

[60] Beniston M, Stephenson DB. Extreme climatic events and their evolution under changing climatic conditions. Global and Planetary Change. 2004;**44**:1-9. DOI: 10.1016/j.gloplacha.2004.06.001

[61] Thomas CD, Cameron A, Green RE, Bakkenes M, Beaumont LJ, Collingham YC, et al. Extinction risk from climate change. Nature. 2004;**427**:145-148. DOI: 10.1038/nature02121

[62] Parmesan C, Root TL, Willig MR. Impacts of extreme weather and climate on terrestrial biota. Bulletin of the American Meteorological Society. 2000;**81**:443-450. DOI: 10.1175/1520-0477(2000)081<0443:IOEWAC>2.3.CO;2

[63] F.A.O. Food and Agriculture Organization. National Institute of Health and National Food Nutrition Survey. Rome: Islamabad, Food balance sheet; 2015

[64] Lesk C, Rowhani P, Ramankutty N. Influence of extreme weather disasters on global crop production. Nature. 2016;**529**:84-87. DOI: 10.1038/nature16467

[65] Maarif S. Meningkatkan kapasitas masyarakat dalam mengatasi risiko bencana kekeringan. Jurnal Sains dan Teknologi Indonesia. 2013;**13**:65-73. DOI: 10.29122/jsti.v13i2.886

[66] Bonan GB, Doney SC. Climate, ecosystems, and planetary futures: The challenge to predict life in earth system models. Science. 2018;**359**(6375):eaam8328

[67] Pratiwi EPA, Ramadhani EL, Nurrochmad F, Legono D. The impacts of flood and drought on food security in Central Java. Journal of the Civil Engineering Forum. 2020;**6**:69. DOI: 10.22146/jcef.51872

[68] Bogati K, Walczak M. The impact of drought stress on soil microbial community. Enzyme Activities and Plants. Agronomy. 2022;**12**:189

[69] Challabathula D, Analin B, Mohanan A, Bakka K. Differential modulation of photosynthesis, R.O.S. and antioxidant enzyme activities in stress-sensitive and -tolerant rice cultivars during salinity and drought upon restriction of COX and AOX pathways of mitochondrial oxidative electron transport. Journal of Plant Physiology. 2022;**268**:153583. DOI: 10.1016/j.jplph.2021.153583

[70] Li Z, Su X, Chen Y, Fan X, He L, Guo J, et al. Melatonin improves drought resistance in maize seedlings by enhancing the antioxidant system and regulating abscisic acid metabolism to maintain stomatal opening under PEG-induced drought. Journal of Plant Biology. 2021;**64**:299-312. DOI: 10.1007/s12374-021-09297-3

[71] Bulgari R, Cocetta G, Trivellini A, Vernieri P, Ferrante A. Biostimulants and crop responses: A review. Biological Agriculture and Horticulture. 2015;**31**:1-17

[72] Rehim A, Amjad Bashir M, Raza Q-U-A, Gallagher K, Berlyn GP. Yield enhancement of biostimulants, vitamin B12, and CoQ10 compared to inorganic fertilizer in radish. Agronomy. 2021;**11**:697. DOI: 10.3390/agronomy11040697

[73] Taha RS, Alharby HF, Bamagoos AA, Medani RA, Rady MM. Elevating tolerance of drought stress in Ocimum basilicum using pollen grains extract; a natural biostimulant by regulation of plant performance and antioxidant defense system. South African Journal of Botany. 2020;**128**:42-53. DOI: 10.1016/j.sajb.2019.09.014

[74] Akensous FZ, Anli M, Boutasknit A, Ben-Laouane R, Ait-Rahou Y, Ahmed HB, et al. Boosting date palm (Phoenix dactylifera L.)

growth under drought stress: Effects of innovative biostimulants. Gesunde Pflanz. 2022:1-22. DOI: 10.1007/s10343-022-00651-0

[75] Xu C, Leskovar DI. Effects of a. nodosum seaweed extracts on spinach growth, physiology and nutrition value under drought stress. Science and Horticulture (Amsterdam). 2015;**183**:39-47. DOI: 10.1016/J.SCIENTA.2014.12.004

[76] Ziaei M, Pazoki A. Foliar-applied seaweed extract improves yield of common bean (Phaseolus vulgaris L.) cultivars through changes in biochemical and fatty acid profile under irrigation regimes. Journal of Soil Science and Plant Nutrition. 2022;**1**:1-11. DOI: 10.1007/s42729-022-00860-6

[77] Pereira JL, Queiroz RML, Charneau SO, Felix CR, Ricart CAO, Lopes Da Silva F, et al. Analysis of Phaseolus vulgaris response to its association with Trichoderma harzianum (ALL-42) in the presence or absence of the phytopathogenic fungi Rhizoctonia solani and fusarium solani. PLoS One. 2014;**9**:e98234. DOI: 10.1371/journal.pone.0098234

[78] Rouphael Y, Giordano M, Cardarelli M, Cozzolino E, Mori M, Kyriacou MC, et al. Plant-and seaweed-based extracts increase yield but differentially modulate nutritional quality of greenhouse spinach through biostimulant action. Agronomy. 2018;**8**:126. DOI: 10.3390/agronomy8070126

[79] Chrysargyris A, Xylia P, Anastasiou M, Pantelides I, Tzortzakis N. Effects of Ascophyllum nodosum seaweed extracts on lettuce growth, physiology and fresh-cut salad storage under potassium deficiency. Journal of the Science of Food and Agriculture. 2018;**98**:5861-5872. DOI: 10.1002/jsfa.9139

[80] Ertani A, Pizzeghello D, Francioso O, Sambo P, Sanchez-Cortes S, Nardi S. Capsicum chinensis L. growth and nutraceutical properties are enhanced by biostimulants in a long-term period: Chemical and metabolomic approaches. Frontiers in Plant Science. 2014;**5**:375. DOI: 10.3389/fpls.2014.00375

[81] Pourghasemian N, Moradi R, Naghizadeh M, Landberg T. Mitigating drought stress in sesame by foliar application of salicylic acid, beeswax waste and licorice extract. Agricultural Water Management. 2020;**231**:105997. DOI: 10.1016/j.agwat.2019.105997

[82] Campobenedetto C, Agliassa C, Mannino G, Vigliante I, Contartese V, Secchi F, et al. A biostimulant based on seaweed (Ascophyllum nodosum and Laminaria digitata) and yeast extracts mitigates water stress effects on tomato (Solanum lycopersicum L.). Agriculture. 2021;**11**:557. DOI: 10.3390/AGRICULTURE11060557

[83] Zeggaf Tahiri A, Carmi G, Ünlü M. Promising water management strategies for arid and semiarid environments. In: Landscape Architecture - Processes and Practices Towards Sustainable Development. London, UK: IntechOpen; 2021 ISBN 978-1-83968-377-0